TRAVEL CAKE

by GARUHARU

GARUHARU MASTER BOOK SERIES 4

TRAVEL CAKE by GARUHARU

초판 1쇄 발행 2022년 12월 30일
초판 4쇄 발행 2024년 6월 3일

지은이 윤은영
어시스턴트 김현철
영문번역 김예성
펴낸이 박윤선
발행처 (주)더테이블

기획·편집 박윤선
디자인 김보라
사진 박성영
스타일링 이화영
영업·마케팅 김남권, 조용훈, 문성빈
경영지원 김효선, 이정민

주소 경기도 부천시 조마루로385번길 122 삼보테크노타워 2002호
홈페이지 www.icoxpublish.com
쇼핑몰 www.baek2.kr (백두도서쇼핑몰)
인스타그램 @thetable_book
이메일 thetable_book@naver.com
전화 032) 674-5685
팩스 032) 676-5685
등록 2022년 8월 4일 제 386-2022-000050 호
ISBN 979-11-979887-4-5 (13590)

더테이블 THE TABLE

GARUHARU MASTER BOOK SERIES 4

TRAVEL CAKE by GARUHARU

트래블 케이크 바이 가루하루

윤은영 지음

더 테이블
THE: TABLE

PROLOGUE

열다섯. 작은 손으로 빚어낸 밀가루 반죽이 오븐 속에서 근사한 과자로 부풀어 오르는 모습을 지켜보았던 날, 제 마음도 따뜻하고 달콤한 과자처럼 부풀어 올랐습니다. 울퉁불퉁 오트밀 쿠키 한 조각을 맛보고 엄지를 추켜세우며 한껏 웃던 가족들과 친구들의 행복한 표정이 저를 '파티시에'라는 길로 이끌어 주었습니다.

파티시에는 다른 이들에게 달콤한 맛의 휴식과 즐거움을 주는 행복한 직업이지만, 기술을 배우고 능숙해지기까지 긴 시간과 노력이 필요했습니다. 하지만 이 길고 힘든 시간 끝에는 저와 제가 만든 제품들이 함께 성장해 있었습니다.

단지 레시피가 아닌 그 과정에서 겪었던 많은 시행착오와 실패를 경험하며 터득한 포인트와 팁, 도구 활용법은 물론 제조 공정에서의 잦은 실수를 줄여주는 방법들을 이 책에 담았습니다. 또한 해외 마스터 클래스를 진행하며 만난 다양한 문화권의 훌륭한 셰프들과 새로운 식재료들로부터 받은 영감의 결과물을 오롯이 담고자 노력했습니다.

이 책이 누군가에게 새로운 영감을 주는 도구로 사용되길 바랍니다. 이 책의 레시피를 토대로 여러분의 주변에서 구할 수 있는 재료를 활용해 다양한 시도를 해보셨으면 좋겠습니다. 제가 늘 작업대 위에서 기존의 제품과 새로운 재료의 조합을 고민하듯, 여러분의 작업실에서도 이러한 고민과 새로운 시도가 계속되길 희망하며, 그 과정에서 저의 책이 작은 도움이 되었으면 합니다.

끝으로 이 책을 펴내는 데 도움을 주신 많은 분들과 아낌없는 지원을 해주신 (주)더테이블 관계자 분들께 고마운 마음을 전합니다.

윤은영

Fifteen. The day I watched the dough I made with my small hands rise up into gorgeous cookies, my heart billowed just like the warm and sweet cookie. The happy faces of family and friends who tasted a piece of the lumpy oatmeal cookie, raising their thumbs up with big loving laughers, led me to the path of becoming a 'pâtissier'.

Being a pâtissier is blissful job, giving others sweet taste of relaxation and delightfulness, but it took a long time and effort to learn the skills and become proficient. However, at the end of this long and challenging time, I and the products I made have been growing together.

This book is not just a collection of recipes, but also reflects a lot of trials and errors that we have experienced along the way, and tried to cover the points and tips we've learned, including how to utilize tools, as well as ways to minimize frequent mistakes during process of making. I also tried to capture the results of inspirations from the talented chefs we've met from various cultures and new ingredients we came across while conducting the overseas master classes.

I hope this book will serve as a new inspiration for someone. Please make variety of attempts with ingredients available near you by utilizing the recipes in this book. Just as I always contemplate the combination of existing products and new ingredients on my workstation, I hope these thoughts and new attempts will continue in your studio, and wish my book will be of little help along the way.

And last but not least, I would like to thank many people who helped me publish this book, and to the officials of THETABLE, Inc. publishers for their generous support.

Yun Eunyoung

CONTENTS

Preparation

INGREDIENTS • 재료 • P.14

BUTTER • 버터 • P.16

SUGAR • 설탕 • P.18

BAKING POWDER AND BAKING SODA • 베이킹파우더와 베이킹소다 • P.20

INGREDIENTS FOR GLUTEN-FREE RECIPES • 글루텐 프리 레시피를 위한 재료들 • P.22

TOOLS • 도구 • P.26

POUND CAKE MOLD • 파운드케이크 몰드 • P.30

CONVECTION OVEN AND DECK OVEN • 컨벡션 오븐과 데크 오븐 • P.32

Basic

30 °B (BAUME) SYRUP • 30보메 시럽 • P.36

GELATIN MASS • 젤라틴매스 • P.38

CHOCOLATE TEMPERING • 초콜릿 템퍼링(접종법) • P.40

Travel Cake

1
MARBLE POUND CAKE • 마블 파운드 케이크 • P.44

2
MATCHA POUND CAKE • 말차 파운드 케이크 • P.58

3
MONT BLANC POUND CAKE • 몽블랑 파운드 케이크 • P.68

4
YUJA POUND CAKE • 유자 파운드 케이크 • P.76

5
FIG POUND CAKE • 무화과 파운드 케이크 • P.84

6
PISTACHIO & GRAPEFRUIT CAKE • 피스타치오 & 자몽 케이크 • P.96

7
CARROT & ORANGE CAKE • 당근 & 오렌지 케이크 • P.104

8
EARL GREY CAKE • 얼그레이 케이크 • P.114

9
CACAO CAKE • 카카오 케이크 • P.128

10
COFFEE BEAN CAKE • 커피 빈 케이크 • P.138

11
CHESTNUT CAKE • 밤 케이크 • P.146

12
CARAMEL PECAN FINANCIER • 캐러멜 피칸 피낭시에 • P.156

13
CHOCOLATE & RASPBERRY FINANCIER • 초콜릿 & 라즈베리 피낭시에 • P.162

14
PISTACHIO & ORANGE FINANCIER • 피스타치오 & 오렌지 피낭시에 • P.168

15
BLUE CHEESE FINANCIER • 블루치즈 피낭시에 • P.174

16
SMOKED VANILLA MADELEINE • 스모크 바닐라 마들렌 • P.182

17
APPLE CRUMBLE MADELEINE • 애플 크럼블 마들렌 • P.188

18
KAFFIR LIME & BASIL MADELEINE • 카피르 라임 & 바질 마들렌 • P.194

19
RASPBERRY MADELEINE • 라즈베리 마들렌 • P.200

20
BURNT VANILLA CANNELE • 번트 바닐라 까눌레 • P.210

21
ROASTED CORN CANNELE • 구운 옥수수 까눌레 • P.220

22
BASIL & RASPBERRY CANNELE • 바질 & 라즈베리 까눌레 • P.228

23
ORANGE & GINGER CANNELE • 오렌지 & 생강 까눌레 • P.236

24
WHITE SOYBEAN PASTE COOKIE • 백된장 쿠키 • P.246

25
DOUBLE CHOCOLATE COOKIE • 더블 초콜릿 쿠키 • P.252

26
VEGAN NUTS COOKIE • 비건 넛츠 쿠키 • P.256

27
EARL GREY COOKIE • 얼그레이 쿠키 • P.260

28
LEMON & BASIL COOKIE • 레몬 & 바질 쿠키 • P.266

29
SEAWEED COOKIE • 김태 쿠키 • P.272

30
CINNAMON & PECAN COOKIE • 시나몬 & 피칸 쿠키 • P.276

31
VEGAN COCO • 비건 코코 • P.282

32

BERRY BERRY MONAKA FLORENTINE • 베리베리 모나카 플로랑탱 • P.288

33

SALTED MONAKA FLORENTINE • 솔티드 모나카 플로랑탱 • P.294

34

SESAME MONAKA FLORENTINE • 깨 모나카 플로랑탱 • P.298

35

COCONUT MONAKA FLORENTINE • 코코넛 모나카 플로랑탱 • P.302

36

RASPBERRY SAND COOKIE • 라즈베리 샌드 쿠키 • P.308

37

SPICED SAND COOKIE • 스파이스 샌드 쿠키 • P.318

38

PISTACHIO SAND COOKIE • 피스타치오 샌드 쿠키 • P.324

39

SALTED BUTTER CARAMEL COOKIE • 솔티드 버터 캐러멜 쿠키 • P.334

40

PRALINE COOKIE • 프랄리네 쿠키 • P.342

41

100% CHOCOLATE COOKIE • 100% 초콜릿 쿠키 • P.350

Preparation

INGREDIENTS

BUTTER

SUGAR

BAKING POWDER AND BAKING SODA

INGREDIENTS FOR GLUTEN-FREE RECIPES

TOOLS

POUND CAKE MOLD

CONVECTION OVEN AND DECK OVEN

재료 Ingredients
소금 (SALT)
달걀 (EGGS)
우유 (MILK)
Le Paludier de Guérande
Fleur de Sel
생크림 (FRESH CREAM)
버터 (BUTTER)
바닐라빈 (VANILLA BEAN)

달걀
포장지에 적힌 산란일자와 포장일을 확인하여 신선한 달걀을 구입하는 것이 중요합니다. 달걀을 사용하기 전에는 물로 깨끗이 씻어 표면에 묻은 이물질을 제거한 후 사용하는 것을 권장합니다. 단, 물로 씻은 달걀을 보관하는 경우에는 세균이나 바이러스 균이 침투할 수 있으므로 필요한 만큼만 세척하여 사용하는 것이 좋습니다.

우유
우유는 제과에 있어 제품의 영양가를 높여줍니다. 또한 단백질과 유당을 함유해 메일라드 반응을 촉진시켜 과자 껍질의 색을 좋게 만듭니다. 고지방 우유, 저지방 우유, 칼슘 강화 우유처럼 특정 영양소를 강화한 우유, 유당을 소화하지 못하는 사람을 위한 락토프리 우유 등 다양한 제품이 있습니다. 이 책에서 사용한 우유는 모두 일반 우유입니다.

생크림
생크림은 식물성 기름이 첨가되지 않은 신선한 것을 구입해 유효기간 내에 사용하는 것이 좋습니다. 식물성 기름을 혼합한 가공 생크림은 작업성을 높여주고 보관 기간이 길지만 입안에서 잘 녹지 않고 풍미가 떨어지기 때문에 권장하지 않습니다. 생크림은 유지방 함량 20% 정도의 저지방 생크림부터 유지방 함량 45% 전후의 고지방 생크림까지 매우 다양합니다. 이책에서는 유지방 함량 38%의 동물성 생크림을 사용하였습니다.

소금
이 책에서 사용한 소금은 프랑스 서부, 대서양을 마주보는 게랑드 일대에서 생산되는 소금입니다. 염전 위에 꽃처럼 피어난 소금 결정을 하나하나 손으로 수확하여 '소금의 꽃(플뢰르 드 셀 fleur de sel)'이라 불립니다. 짠맛 뿐만 아니라 감칠맛까지 느낄 수 있으며 캐러멜과 같은 단맛과도 잘 어우러집니다.

바닐라빈
바닐라빈은 우유, 생크림, 버터와 같은 유제품 향의 조화를 높여주며 달걀 특유의 비릿함을 잡아줍니다. 긁어낸 바닐라빈은 크림이나 반죽에 넣어 향을 내고, 사용하고 남은 껍질은 설탕에 넣어 두었다가 설탕과 함께 푸드프로세서에 갈아 바닐라슈거를 만들어 사용할 수 있습니다.

버터
제과의 기본 재료인 버터는 냉동고나 냉장고에 넣어 차가운 상태, 실온에 꺼내두어 말랑한 상태, 전자레인지에 녹여 완전히 풀어진 상태 세 가지 형태로 사용합니다. 버터는 유지방 함량이 80% 이상인 퓨어버터를 사용하는 것이 좋습니다. 산화되기 쉬운 퓨어버터는 신선한 것을 구입해 빠른 시일 안에 사용하고, 오래 두고 사용해야 할 경우에는 냄새를 흡수하지 않도록 잘 밀폐하여 냉동 보관합니다. 이 책에서 사용한 버터는 제조 공정 중 유산균을 첨가해 숙성한 프랑스 천연 발효 버터입니다. 느끼함이 없고 유지방이 응축된 고소하고 담백한 맛을 느낄 수 있습니다.

Eggs
It is important to purchase fresh eggs by checking the date eggs were laid* and the packaging date printed on the carton. It is recommended to wash them thoroughly with water before using them to remove foreign substances on the surface. However, there is a possibility that bacteria or viruses may penetrate while storing eggs washed with water, so it's better to wash them only as needed.

* May not be indicated in some countries.

Milk
Milk improves the nutritional value of the baked goods in confectionery. It also contains protein and lactose, which promotes the Maillard reaction that helps to make an appetizing outer color of the confectionery goods. There are various kinds of milk, which include high-fat milk, low-fat milk, and milk with certain nutrients strengthened such as calcium-fortified milk, and lactofree milk for those who cannot digest lactose. The milk used in this book is regular whole milk.

Fresh cream (Heavy cream, Whipping cream)
It is recommended to purchase fresh cream that does not contain vegetable oils and use it within the expiration date. Processed whipping cream blended with vegetable oils increases workability and has a long shelf life; however, it is not recommended because it does not melt nicely in the mouth and has poor flavor. There are varieties of fresh creams ranging from low-fat fresh cream with a milk fat content of about 20% to high-fat fresh cream with a milk fat content of around 45%. In this book, we used fresh dairy cream containing 38% of milk fat.

Salt
The salt used in this book is produced in the region of Guerande, facing the Atlantic Ocean in western France. Each salt crystal blooming like flowers on the salt field are harvested by hand is called 'flower of salt (fleur de del).' You can feel not only its saltiness but also savory taste, and pairs well with sweet flavors such as caramels.

Vanilla beans
Vanilla bean enhances the harmony of dairy flavors such as milk, fresh cream, and butter and helps reduce the peculiar taste of eggs some may find fishy. The scraped vanilla bean seeds can be added to creams or batters to enhance flavor, and the remaining skin can be stored with sugar, which can be ground all together with a food processor to make vanilla sugar.

Butter
Butter, which is the basis of confectionery ingredients, is used in three forms: cold state by keeping in a freezer or refrigerator, soft state by keeping in ambient temperature, and completely dissolved state by melting in a microwave. As for the butter, pure butter containing more than 80% of milk fat would be recommended to use. Pure butter should be purchased fresh and used as soon as possible because it's easy to oxidize. If it needs to be stored for a long time, it should be sealed tight, not to absorb any odor, and kept frozen. The butter used for this book is naturally fermented butter from France, aged by adding lactic acid bacteria during the manufacturing process. It doesn't taste greasy and gives a savory and clean flavor from concentrated milk fat.

버터 Butter

포마드 상태의 버터
(POMMADE (SOFTENED) BUTTER)
녹인 버터
(MELTED BUTTER)
단단한 상태의 버터
(HARDENED BUTTER)

제과에서 사용하는 버터는 크게 냉장고에 보관한 단단한 상태의 차가운 버터, 실온에 두어 말랑해진 포마드 상태의 버터, 열을 가해 액체 상태로 녹인 버터 총 세 가지로 나뉩니다.
이 책에서는 이 세 가지 상태의 버터 외에 정제된 버터도 사용합니다. 일반적인 버터의 녹는점이 약 28~35℃인 반면 이 책에서 사용된 정제버터는 약 17℃의 낮은 융점을 가지고 있습니다. 이는 일반적인 실내 온도보다 낮은 온도라고 볼 수 있습니다. 이로 인해 정제버터를 사용한 가루하루의 마들렌은 버터의 풍미를 잃지 않으면서 실온에서도 부드러운 식감을 느낄 수 있습니다.

* 이 책에서 사용한 정제버터는 '콜만 액상버터'입니다.

The butter used in confectionery is largely divided into three types: cold butter in a hard state stored in the refrigerator, butter in a pommade (softened) state that has softened at room temperature, and butter melted in a liquid form by applying heat.
In addition to butter in these three states, clarified butter is also used in this book. While regular butter has a melting point of about 28~35°C, the clarified butter used in this book has a low melting point of about 17°C. This can be considered as a lower than the normal room temperature. As a result, Garuharu's madeleines made with clarified butter have a soft texture even at room temperature without losing the flavor of the butter.

* The clarified butter used in this book is 'Corman Liquid Clarified Butter.'

버터 Butter		
근원 Origin	녹는점 Melting point	발연점 Smoking point
동물 Animal	28~35℃	150~177℃
성질 Characteristics		
크리밍성 Creaming ability (23~25℃)	가소성 Plasticity (15~18℃)	쇼트닝성 Shortening property (20~23℃)
버터에 공기를 포집하는 성질 예: 크림법, 버터크림 Traps air in the butter i.e.) creaming method, buttercream	버터에 물리적인 힘이 가해질 때 자유자재로 형태가 변하는 성질 예: 푀이타주, 데니시 페이스트리 Freely changes its shape when physical force is applied to butter i.e.) feuilletage, Danish pastry	버터 입자들이 반죽 속에서 고르게 분산되어 밀가루 글루텐 형성을 억제하고 전분의 결합을 방해해 바삭한 식감으로 만들어주는 성질 예: 사블레, 크럼블 Butter particles are evenly dispersed in the dough, suppressing the formation of gluten and preventing the bonding of starch to make it crispy. i.e.) sable, crumble
녹은 버터는 이 세 가지 성질을 모두 잃습니다. Melted butter loses all three characteristics.		

설탕 Sugar

무스코바도 설탕
(MUSCOVADO SUGAR)

당밀 (MOLASSES)

원당
(RAW SUGAR)

백설탕
(WHITE SUGAR)

설탕의 가공 과정

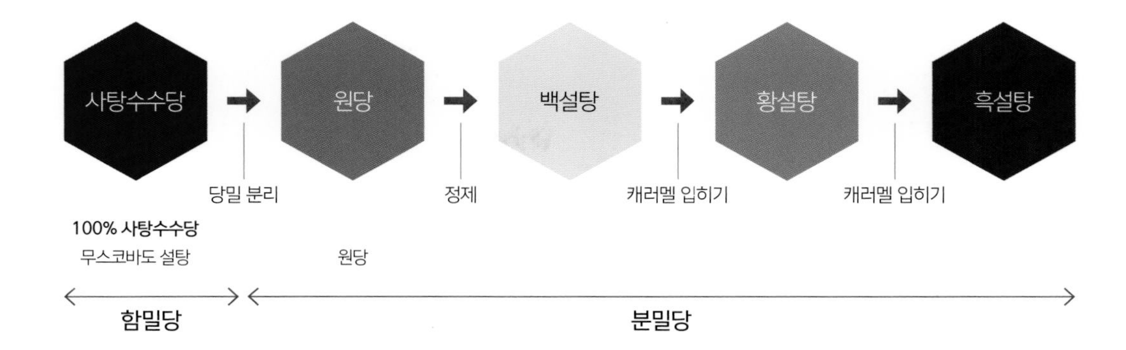

함밀당 (100% 사탕수수당)

함밀당은 사탕수수의 즙에 알칼리를 넣어 끓인 후 수분을 증발시켜 만든 결정으로 무스코바도 설탕이 여기에 속합니다. 당밀을 분리하지 않아 어두운 색을 내며 무기질, 비타민 성분이 많이 함유되어 있습니다. 또한 불순물도 함께 섞여 있기 때문에 정제 과정을 거친 백설탕과 다르게 맛과 향이 풍부하며 독특합니다.

분밀당 (원당, 원료당)

원심분리기를 이용해 함밀당에서 당밀을 분리한 황갈색의 설탕입니다.

백설탕

분밀당을 정제, 가공해 만든 무색 무취의 백색의 설탕입니다.

황설탕 & 흑설탕

정제, 가공 과정을 거친 백설탕에 캐러멜 색소를 입힌 설탕입니다.

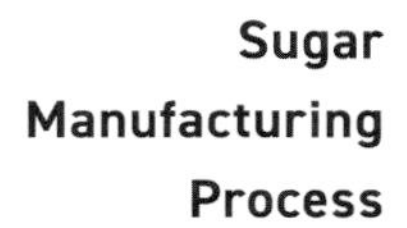

Sugar Manufacturing Process

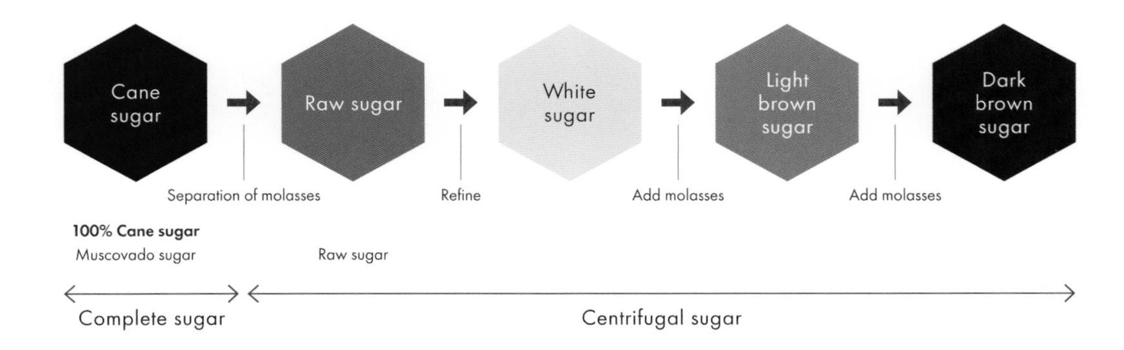

Complete Sugar (100% Cane sugar)

Complete sugar is a crystal made by boiling the juice of sugar cane and alkali and then evaporating the moisture. Muscovado sugar belongs to this category. It has dark color because the molasses is not separated and contains many minerals and vitamins. Also, because it is mixed with impurities, it has a rich and unique taste and aroma, unlike white sugar, which has been refined.

Centrifugal Sugar (Raw sugar)

It is light-brown sugar obtained by separating molasses from complete sugar using a centrifuge machine.

White Sugar (Granulated sugar)

It's colorless and odorless white sugar made by refining and processing raw sugar.

Light Brown Sugar & Dark Brown Sugar

It refers to refined and processed white sugar coated with caramel color or molasses.

베이킹파우더와 베이킹소다
Baking Powder and Baking Soda

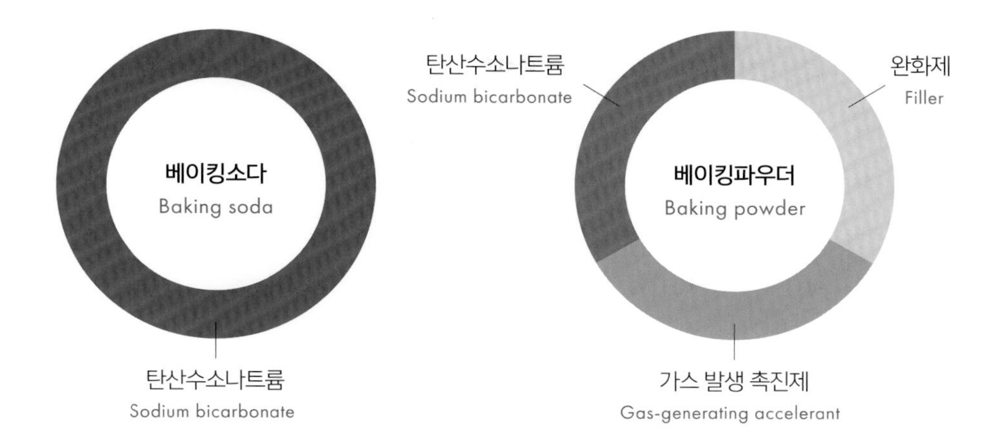

베이킹소다

베이킹소다는 탄산수소나트륨만으로 구성된 제품입니다. 수분에 쉽게 녹으며 산과 열에 의해 탄산나트륨, 물, 이산화탄소로 분해되며 이 이산화탄소가 가스를 발생시켜 반죽의 부피를 팽창시키게 됩니다. 베이킹소다를 쿠키 반죽에 넣게 되면 반죽을 퍼지게 하면서 팽창시켜줍니다. 단, 반죽 속에 충분한 산이 없는 경우 분해되지 못하고 반죽 속에 남아 특유의 쓴맛을 유발시고 제품의 색을 노랗게 변색시킬 수 있습니다.

베이킹파우더

탄산수소나트륨에 가스 발생 촉진제(산성제)와 완화제(전분)을 혼합한 제품입니다. 가스 발생 촉진제가 탄산수소나트륨의 반응을 촉진시키며 알칼리를 중화시켜 제품에 쓴맛이 남지 않으며 변색되지도 않습니다. 완화제는 개봉된 제품이 공기 중의 수증기에 의해 반응하는 것을 막아주는 역할을 합니다.

Baking Soda

Baking soda is a product consisting of only sodium bicarbonate. It dissolves easily in water and decomposes into sodium carbonate, water, and carbon dioxide by acid and heat, and the carbon dioxide generates gas and expands the volume of the batter or dough. When baking soda is added to the cookie dough, it expands as it enables it to spread. However, if there is not enough acid in the dough, it cannot be decomposed and remains in the dough, causing a peculiar bitterness and turning the color of the product yellow.

Baking Powder

It is a mixture of sodium bicarbonate with a gas-generating accelerant (acid) and a filler (starch). The gas-generating accelerant promotes the reaction of sodium bicarbonate and neutralizes the alkali, leaving no bitter taste or discoloration in the product. The filler prevents the opened product from reacting with water vapor in the air.

글루텐 프리 레시피를 위한 재료들

Ingredients for Gluten-free Recipes

디저트 샵을 운영하다보면 알레르기로 인해 글루텐 섭취가 어려운 분들을 종종 만나게 됩니다. 이런 분들을 보며 팀 가루하루의 호기심과 고민이 시작되었고, 수많은 테스트를 거쳐 지금의 글루텐 프리 레시피를 완성할 수 있었습니다.
이 책에서 소개하는 글루텐 프리 레시피를 활용하면 글루텐 알레르기가 있으신 분들은 물론 건강상의 이유로 글루텐 섭취를 제한하고 싶은 분들까지도 맛있고 건강한 가루하루의 디저트를 즐기실 수 있습니다.

○ 쌀가루

시판되는 쌀가루의 종류

쌀가루는 침지(불림) 과정의 유무에 따라 '건식쌀가루'와 '습식쌀가루'로 구분되며 각각의 제조 과정은 아래와 같습니다.

• 건식쌀가루의 제조 과정

세척	탈수	제분	건조

• 습식쌀가루의 제조 과정

세척	침지(불림)	탈수	제분	건조

침지 과정은 쌀 원물을 물에 불려 전분 입자 사이의 간격을 벌어지게 해 스펀지와 같은 조직을 만드는 과정입니다. 상온에서 유통되는 습식 쌀가루의 경우 이렇게 쌀 전분 입자를 벌려 놓은 상태에서 건조시키므로(상온 유통 과정 중 미생물이 번식하는 것을 방지하기 위해) 순간적으로 많은 양의 수분을 흡수할 수 있게 됩니다.
제과에서 밀가루 대체 재료로 쌀가루를 사용할 경우에는 수분 흡수와 보유력이 좋은 습식쌀가루를 사용하는 것이 적합합니다. 건식쌀가루의 경우 반죽 과정에서 쌀 전분이 충분한 수분을 흡수할 수 없어 반죽이 질어지고 퍼지는 요인이 되기 때문입니다. 또한 입자가 고울수록 부드러운 식감을 주기 때문에 기호나 용도에 따라 쌀가루의 입자 크기를 선택해 사용할 수 있습니다. 시판되고 있는 박력쌀가루의 경우 부드러운 식감을 위해 밀가루와 비슷한 입자로 제분한 것입니다.

When you run a dessert shop, you will often meet people who have difficulty ingesting gluten due to allergies. As I watched these people, Team GARUHARU's curiosity and concerns began, and after numerous tests, we were able to complete the current gluten-free recipe.
By using the gluten-free recipe introduced in this book, you can enjoy GARUHARU's delicious and healthy tarte, not only for those with gluten allergies but also for those who want to limit gluten intake for health reasons.
We hope you will enjoy the various styles of desserts by choosing recipes presented in this book.

○ Rice flour

Types of rice flour available in the market

Rice flour is divided into 'dry(-milled) rice flour' and 'wet(-milled) rice flour' depending on whether or not it went through the process of immersion (soaking), and each manufacturing process is as follows.

• Manufacturing process of dry rice flour

Wash	Dehydration	Milling	Dry

• Manufacturing process of wet rice flour

Wash	Immersion (soaking)	Dehydration	Milling	Dry

The immersion is a process in which raw rice grains are soaked in water to open the gaps between starch particles to create a sponge-like tissue. In the case of the wet rice flours distributed at ambient temperature, rice starch particles are dried in an open state (to prevent microorganisms from multiplying during the distribution process in ambient temperature) so that a large amount of moisture can be absorbed instantaneously.
When using rice flour to substitute wheat flour in confectionery, it is suitable to use wet rice flour, which has good moisture absorption and retention. This is because, when dry rice flour is used, the rice starch cannot absorb enough moisture during the kneading process, which can cause the dough to become thin and run easily. Additionally, the finer the particles, the softer the texture, so you can select the particle size of rice flour and use it according to your taste or purpose. The commercially available soft rice flour is ground into particles similar to wheat flour for a smooth texture.

1
2
3
4
5-1
5-2

쌀가루 만들기

1. 쌀을 흐르는 물에 3~4회 깨끗이 세척한다.
2. 냉장고에서 12시간 정도 불린다.
3. 체에 받쳐 30분간 탈수한다.
4. 45℃에서 10시간 정도 건조시킨다.
5. 푸드프로세서에 넣고 고운 가루 상태가 될 때까지 갈아준다.

How to make rice flour

1. Wash rice 3~4 times under running water.
2. Let it soak in refrigerator for about 12 hours.
3. Strain and dehydrate for 30 minutes.
4. Dry at 45˚C for 10 hours.
5. Grind in food processor until it turns into fine powder.

도구 Tools

푸드프로세서
(FOOD PROCESSOR)

짤주머니 (PIPING BAG)

저울 (SCALE)

깍지 (TIPS)

핸드블렌더
(HAND BLENDER)

스크래퍼
(SCRAPER)

믹싱볼 (MIXING BOWLS)

스패출러 (SPATULA)

저울

베이킹에 있어 정확한 계량은 매우 중요합니다. 1g 차이로도 맛이 달라지거나 제품이 아예 만들어지지 않을 수도 있기 때문입니다. 따라서 눈금저울보다는 1g 단위로 표시되는 전자저울을 사용하는 것이 좋습니다. 재료를 담는 볼을 올리고 0을 맞춰 0점 조절을 한 다음 계량을 시작합니다. 1g보다 더 적은 양을 계량한다면 1g에서 1/2, 1/4과 같이 줄여나가면 됩니다.

푸드프로세서

재료를 잘게 자르고 다지는 용도로 사용합니다.

짤주머니/깍지

반죽이나 크림을 담아 짜거나 모양을 내기 위해 사용하는 도구입니다. 굽는 반죽의 경우 천 재질의 짤주머니를 사용해도 좋지만, 가열하는 과정 없이 바로 먹는 크림 종류의 경우에는 식중독 예방을 위해 일회용 비닐 짤주머니를 사용하고 재사용하지 않는 것이 위생적입니다. 다양한 종류의 깍지를 짤주머니에 끼워 용도에 맞게 사용합니다.

믹싱볼

반죽을 섞거나 거품을 올리는 작업에 사용되는 필수 도구입니다. 스테인리스, 유리, 폴리카보네이트 등 다양한 재질이 있으며 용도에 맞춰 큰 사이즈(지름 25cm)부터 작은 사이즈(17cm)까지 3~4개 정도 여유롭게 준비하는 것이 좋습니다. 스테인리스 재질은 가볍고 열 전달이 빠른 반면, 유리와 폴리카보네이트 재질은 상대적으로 열 전달이 느려 뜨거운 것이나 차가운 것을 보다 오래 유지시켜줍니다.

핸드블렌더

글레이즈, 크림, 가나슈 등을 균일하게 혼합하는 데 유용한 도구입니다.

스패출러

반죽이나 크림을 펼치거나 고르게 정리할 때 사용합니다. 손잡이와 날이 일직선인 것, L자로 굽은 것 두 가지 종류가 있습니다.

스크래퍼

넓은 면적을 이용해 반죽을 이기거나 혼합하는 데 사용하며, 반죽을 떠 담기에도 유용한 도구입니다.

Scale

Accurate scaling is extremely important in baking. It's because the taste may change, or the product may not be made at all, even with a 1 gram of difference. Therefore, it is better to use an electronic scale that displays in 1-gram increments rather than a needle-type scale. Place a bowl for scaling the ingredient on the scale and set to 0 (tare), and start weighing. If you measure less than 1 gram, reduce from 1 gram to 1/2, 1/4, and so on.

Food processor

This is used to chop and grind ingredients.

Piping bags/tips (nozzles)

These are used for piping or shaping batter or cream. Cloth material piping bags may be used for batters that will be baked. But for creams consumed without the cooking process, it is recommended to use disposable plastic piping bags for hygienic reasons and not to re-use them to prevent food poisoning. Use various types of tips with piping bags accordingly.

Mixing bowls

It is an indispensable tool used for mixing doughs or whipping. There are various materials such as stainless steel, glass, polycarbonate, etc. It is recommended to prepare 3~4 bowls, from large size (25cm diameter) to small size (17cm diameter), to use according to the purpose. Stainless steel material is light and transfers heat faster, while glass and polycarbonate materials have relatively slower heat transfer, which helps keep hots or colds longer.

Hand blender (Immersion blender)

A hand blender is a useful tool for mixing glazes, creams, and ganache evenly.

Spatula

It is used to spread or evenly organize cream or batter. There are two types: one that's straight from the handle to the blade and one in which the blade is bent to L-shape.

Scraper

A scraper is used to beat or mix batters and is also a useful tool for transferring batter.

TFX 410
ON / OFF
온도계
(THERMOMETER)
실리콘 주걱 (SILICONE SPATULA)
거품기 (WHISK)
체 (STRAINER)
테프론시트
(TEFLON SHEET)
실리콘매트 (SILICONE MAT)
에어매트 (AIR MAT)

오븐용 페이퍼

오븐용 페이퍼는 유산지, 테프론시트, 실리콘매트 등의 제품을 말하며 철판에 깔아 제품과 철판이 달라붙는 것을 방지하고 구웠을 때 제품의 밑면이 깔끔하게 분리되게 하는 역할을 합니다. 실리콘 페이퍼, 내열 페이퍼, 방수 페이퍼 등의 특수 가공 처리를 한 제품들도 있습니다.

에어매트

그물 모양으로 만들어진 오븐용 실리콘매트입니다. 촘촘하게 구멍이 뚫려 있기 때문에 오븐 안에서 공기의 흐름이 좋고 반죽이 미끄러지지 않아 모양이 그대로 유지됩니다.

온도계

정확한 계량 못지않게 베이킹에서 중요한 것이 바로 온도입니다. 온도계는 주로 반죽이나 크림의 온도를 체크하는 데 사용되며 고온의 시럽을 만들 때에도 사용할 수 있도록 250℃까지 측정이 되는 제품을 선택하는 것이 좋습니다. 표면의 온도를 측정하는 비접촉식 적외선 온도계와 내용물 안에 넣어 온도를 측정하는 접촉식 온도계 두 가지 종류가 있습니다.

실리콘 주걱

볼에 담긴 반죽이나 크림을 섞거나 모으는 데 사용합니다. 힘 없이 휘어지는 것보다는 어느 정도 탄력이 있는 제품이 사용하기 더 편리합니다. 주걱 앞부분부터 손잡이 끝부분까지 홈이 없이 매끄럽게 연결되어 만들어진 주걱이 이물질이 끼지 않아 위생적으로 더 좋습니다.

거품기

재료를 섞어 혼합하거나 공기를 포집할 때 사용하는 도구입니다. 와이어 부분이 유연하면서도 튼튼하며 손으로 잡았을 때 편한 것을 고르는 것이 좋습니다.

체

가루 재료를 곱게 내리거나 섞어 놓은 액체 재료를 거르는 데 사용하는 도구로 다양한 사이즈가 있습니다. 제과에서 가루 재료들은 모두 체에 내린 후 사용하는 것이 기본입니다. 가루 재료를 체에 내리면 뭉쳐 있던 멍울은 풀어지고, 불순물이 제거되며, 체에 내리는 과정에서 가루와 가루 사이에 공기가 혼입되어 다른 재료와 더 잘 섞이게 됩니다.

Papers for oven

Papers for the oven refers to parchment paper, Teflon sheets, and silicone mats. These are to be lined on a baking tray to prevent products from sticking and to help the bottom of the baked goods separate neatly. There are also specially processed products such as silicone paper, heat-resistant paper, and waterproof paper.

Air mat (Perforated silicone mat)

This is a net-shaped silicone mat for ovens. Because of its fine holes, the air in the oven flows through and around, and the dough does not slip on it, which helps to maintain its shape.

Thermometer

Temperature is every bit as important as accurate scaling when it comes to baking. Thermometers are mainly used to check the temperature of dough/batter or cream and check the temperature of high-temperature syrups. Therefore, it is recommended to select a product that can measure up to 250°C. There are two types: a non-contact type infrared thermometer that measures surface temperature and a contact type thermometer that measures the temperature by placing it inside the content.

Silicone Spatulas

These are used to mix or collect batter or cream in the bowl. Spatula with a resilient blade is more convenient to use than the blades that bend easily. The spatula that is smooth all the way, and has no groove from the blade to the tip of the handle, is better in the means of hygiene because no foreign substance will remain in between.

Whisk

This is a tool used to mix ingredients or to collect air. Choosing a whisk with wires that are flexible yet sturdy and comfortable when held by hand is good to use.

Strainer

A strainer is used to sift powder ingredients a fine form or filter mixture of liquid ingredients, and there are various sizes available. In confectionery, it is a ground rule to sieve all powdered ingredients. When the powdered ingredients are sifted, the lumps break up, impurities are removed, and the air gets to mix in between the particles, which helps to better mix with other ingredients.

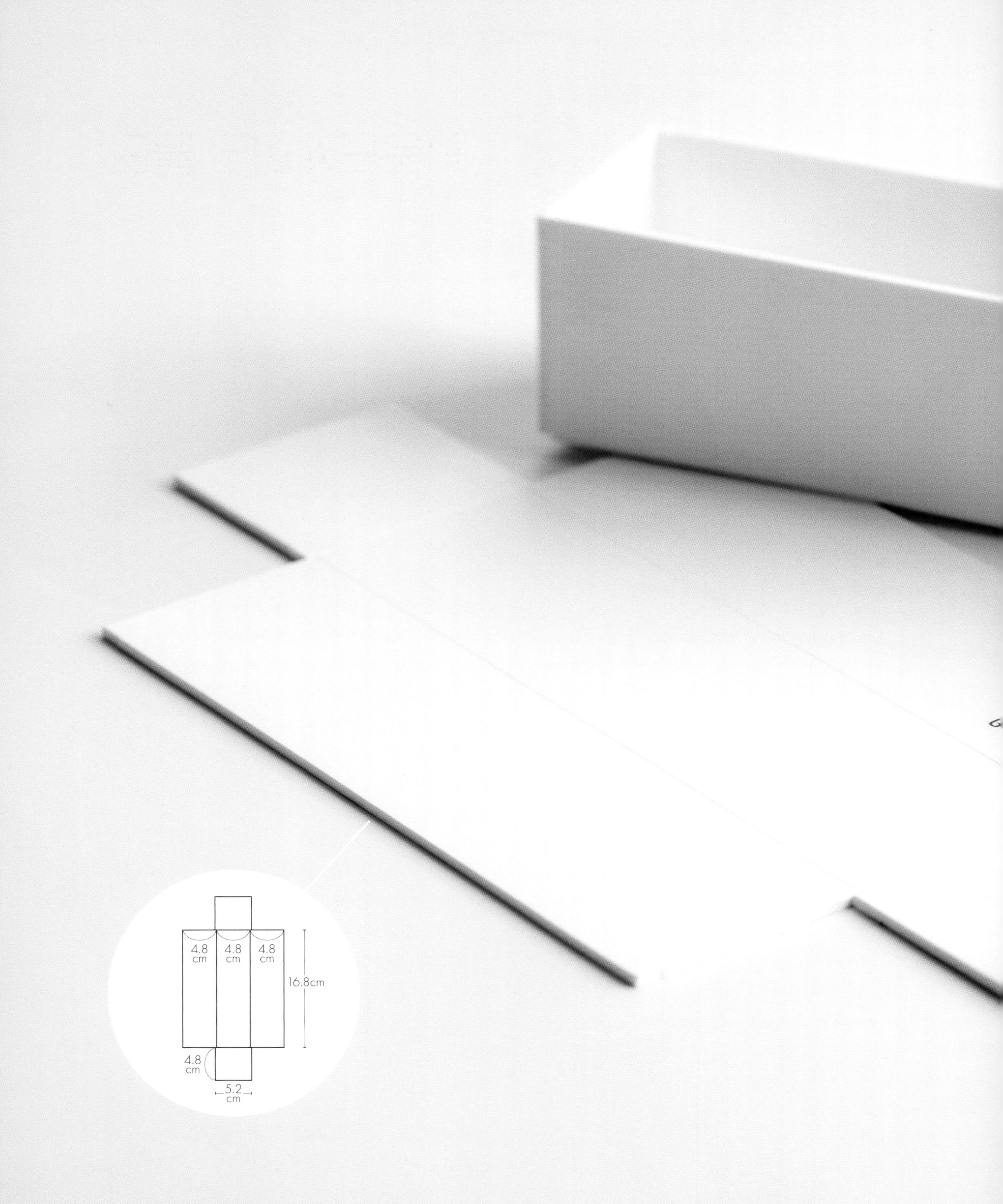
4.8
cm
4.8
cm
4.8
cm
16.8cm
4.8
cm
5.2
cm

파운드케이크 몰드

Pound Cake Mold

이 책에서 소개하는 파운드케이크는 가루하루에서 직접 제작한 폼보드 몰드를 사용해 완성했습니다. 동일한 사이즈로 제작해 책과 같은 모양으로 완성할 수도 있고, 잘 구워진 파운드케이크를 그대로 즐길 수도 있습니다. 좀 더 오래 보관하고 싶은 경우 실온 유지 초콜릿 글레이즈나 폰당과 같은 재료들로 코팅할 수도 있습니다. 다양한 방식으로 파운드 케이크를 즐겨보시기 바랍니다.

The pound cakes presented in this book are made using custom-made foam board molds by GARUHARU. You can make the mold in the same size and shape as the book or enjoy the well-baked pound cakes as it is. If you want to keep them a little longer, you can also coat them with room-temperature chocolate glaze or fondant. Enjoy the pound cakes in a variety of ways.

컨벡션 오븐과 데크 오븐

Convection Oven and Deck Oven

오븐의 종류는 '데크 오븐'과 '컨벡션 오븐'으로 나눌 수 있습니다.
데크 오븐은 오븐의 상부와 하부에 위치한 열선을 통해 열을 전달하며, 컨벡션 오븐은 오븐 내부의 팬이 오븐 안의 열을 순환시키는 방식으로 열을 전달합니다.
컨벡션 오븐의 경우 뜨거운 바람으로 열을 전달하기 때문에 다단 조리가 가능한 장점이 있지만 반죽이 건조해지기 쉽습니다.
따라서 컨벡션 오븐은 바삭한 식감의 쿠키나 파이 등을 굽기에 적합하며, 촉촉한 식감의 제품을 구울 때는 적합하지 않습니다.

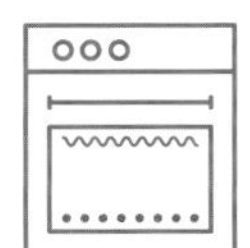

데크 오븐 DECK OVEN

상부와 하부의 열선을 통해 열을 전달

Heat distributed through
the upper and lower heating wires

컨벡션 오븐 CONVECTION OVEN

팬의 바람을 통해 열을 순환

Distributes heat by circulating heated air
with a fan

The types of ovens can be divided into 'deck oven' and 'convection oven.'
The deck oven distributes heat through the heating wires located at the top and bottom of the oven. As for the convection oven, heated air is distributed by the fan inside the oven circulating the heat.
Convection ovens have the advantage of being able to cook multiple trays because it transfers heat by hot air, but the dough may easily dry out. Therefore, a convection oven is ideal for baking crispy textured cookies or pies and not suitable for baking products that need to be moist.

Basic

30 °B (BAUME) SYRUP

GELATIN MASS

CHOCOLATE TEMPERING

30 °B (BAUME) SYRUP

30 °B (BAUME) SYRUP

30보메 시럽

ingredients

물 1000g
설탕 1350g

1000g Water
1350g Sugar

Process

1. 냄비에 물과 설탕을 넣고 설탕이 녹을 때까지 가열한다.
2. 충분히 식으면 밀폐 용기에 담아 냉장 보관하며 사용한다.

1. Heat water and sugar in a pot until sugar dissolves.
2. When the syrup is cooled enough, store in an airtight container and refrigerate to use.

GELATIN MASS
1
2-1
2-2
3

GELATIN MASS

젤라틴매스

응고제의 일종인 젤라틴은 무스나 젤리 등을 굳히는 데 사용되는 재료입니다. 이 책에서는 가루 형태의 젤라틴(200 bloom)을 5배의 물과 혼합해 굳힌 후 사용하였습니다.

Gelatine is a kind of coagulant that is used to harden mousse or jelly. For this book, the powder type of gelatin (200 bloom) is mixed with five times of water and hardened to use (gelatin mass).

ingredients

200 Bloom 가루젤라틴 10g
물 50g

10g Powdered gelatin, 200 bloom
50g Water

Process

1. 가루젤라틴과 물을 1:5 비율로 계량한다.
2. 가루젤라틴과 물을 혼합한다.
3. 젤라틴이 굳으면 필요한 만큼 적당한 크기로 잘라 사용한다. 냉장 상태에서 2주, 냉동 상태에서 3개월 동안 보관하며 사용할 수 있다.

1. Scale powdered gelatin and water in the ratio of 1:5.
2. Mix powdered gelatin and water.
3. When the gelatin is set, cut into moderate size to use. It can be stored for about two weeks in a refrigerator and three months in a freezer.

1

2

3-1

3-2

4

CHOCOLATE TEMPERING

초콜릿 템퍼링(접종법)

접종법은 녹인 초콜릿에 안정된 결정 상태의 초콜릿을 넣어 템퍼링하는 방법입니다. 여러 가지 템퍼링 방법 중 작업 방식이 간편해 대량 작업 시에도 비교적 손쉽게 템퍼링할 수 있습니다.

The seeding method is a technique of tempering, where stable crystallized chocolate is added to melted chocolate. This method is simple among various methods of tempering, which also makes it relatively easy to work with a large amount of chocolate.

ingredients

커버추어 초콜릿

couverture chocolate

Process

1. 커버추어 초콜릿을 폴리카보네이트 볼에 담고 녹여준다. (다크초콜릿 55~58℃, 밀크초콜릿 45~48℃, 화이트초콜릿 45~48℃)
2. 녹인 초콜릿 양의 30% 정도의 초콜릿을 1에 넣고 골고루 저어준 후 1분간 그대로 둔다.
3. 핸드블렌더를 이용해 초콜릿 덩어리가 남지 않도록 균일하게 믹싱하며 온도를 낮춰준다. (다크초콜릿 31~32℃, 밀크초콜릿 29~30℃, 화이트초콜릿 28~29℃)
   * 이때 빠른 속도로 장시간 믹싱하면 마찰열에 의해 온도가 과도하게 올라갈 수 있으므로 주의한다.
4. 스패출러 또는 나이프 끝부분에 템퍼링 테스트를 한 후 사용한다.
   * 23~24℃를 유지한 작업실에서 5분 이내에 얼룩 없이 매끄럽게 초콜릿이 굳으면 템퍼링이 잘된 상태이다.

1. Melt couverture chocolate in a polycarbonate bowl. (Dark chocolate 55~58°C, milk chocolate 45~48°C, white chocolate 45~48°C)
2. Stir in evenly about 30% of the melted chocolate to 1, and let stand for 1 minute.
3. Reduce the temperature of chocolate using a hand blender so that it's mixed evenly and no chocolate chunks are left. (Dark chocolate 31~32°C, milk chocolate 29~30°C, white chocolate 28~29°C)
   * Keep in mind when mixing at high speed for a long time, the temperature can excessively increase due to heat caused by friction.
4. Test on the tip of a spatula or knife before use.
   * If the chocolate hardens smoothly within 5 minutes in a studio that maintained 23~24°C, the tempering is done properly.

Travel Cake

MARBLE POUND CAKE

MATCHA POUND CAKE

MONT BLANC POUND CAKE

YUJA POUND CAKE

FIG POUND CAKE

PISTACHIO & GRAPEFRUIT CAKE

CARROT & ORANGE CAKE

EARL GREY CAKE

CACAO CAKE

COFFEE BEAN CAKE

CHESTNUT CAKE

CARAMEL PECAN FINANCIER

CHOCOLATE & RASPBERRY FINANCIER

PISTACHIO & ORANGE FINANCIER

BLUE CHEESE FINANCIER

SMOKED VANILLA MADELEINE

APPLE CRUMBLE MADELEINE

KAFFIR LIME & BASIL MADELEINE

RASPBERRY MADELEINE

BURNT VANILLA CANNELE

ROASTED CORN CANNELE

BASIL & RASPBERRY CANNELE

ORANGE & GINGER CANNELE

WHITE SOYBEAN PASTE COOKIE

DOUBLE CHOCOLATE COOKIE

VEGAN NUTS COOKIE

EARL GREY COOKIE

LEMON & BASIL COOKIE

SEAWEED COOKIE

CINNAMON & PECAN COOKIE

VEGAN COCO

BERRY BERRY MONAKA FLORENTINE

SALTED MONAKA FLORENTINE

SESAME MONAKA FLORENTINE

COCONUT MONAKA FLORENTINE

RASPBERRY SAND COOKIE

SPICED SAND COOKIE

PISTACHIO SAND COOKIE

SALTED BUTTER CARAMEL COOKIE

PRALINE COOKIE

100% CHOCOLATE COOKIE

1 MARBLE POUND CAKE

마블 파운드 케이크

ingredients - 16.5 × 4.5 × 4.5cm, 3 cakes

마블 파운드 케이크

달걀노른자 142g
설탕 168g
플레인요거트 114g
박력분 147g
베이킹파우더 4.8g
포도씨유 15g
버터 72g
카카오파우더 22g

골드럼 시럽

30보메 시럽 200g
물 30g
골드럼 115g
(MONARCH)

기타

다크초콜릿
(CARAIBE 66%)
구운 헤이즐넛

비터 초콜릿 가나슈

생크림 199g
전화당 45g
다크초콜릿 199g
(CARAIBE 66%)
버터 45g
골드럼 12g
(MONARCH)

헤이즐넛 프랄리네

헤이즐넛 250g
바닐라빈 2개
설탕 250g
물 58g
게랑드소금 2.5g

MARBLE POUND CAKE

142g Egg yolks
168g Sugar
114g Plain yogurt
147g Cake flour
4.8g Baking powder
15g Grapeseed oil
72g Butter
22g Cacao powder

GOLD RUM SYRUP

200g 30°B syrup
30g Water
115g Gold rum
(MONARCH)

OTHER

Dark chocolate
(CARAIBE 66%)
Roasted hazelnuts

BITTER CHOCOLATE GANACHE

199g Heavy cream
45g Inverted sugar
199g Dark chocolate
(CARAIBE 66%)
45g Butter
12g Gold rum
(MONARCH)

HAZELNUT PRALINE

250g Hazelnuts
2 Vanilla beans
250g Sugar
58g Water
2.5g Guérande salt

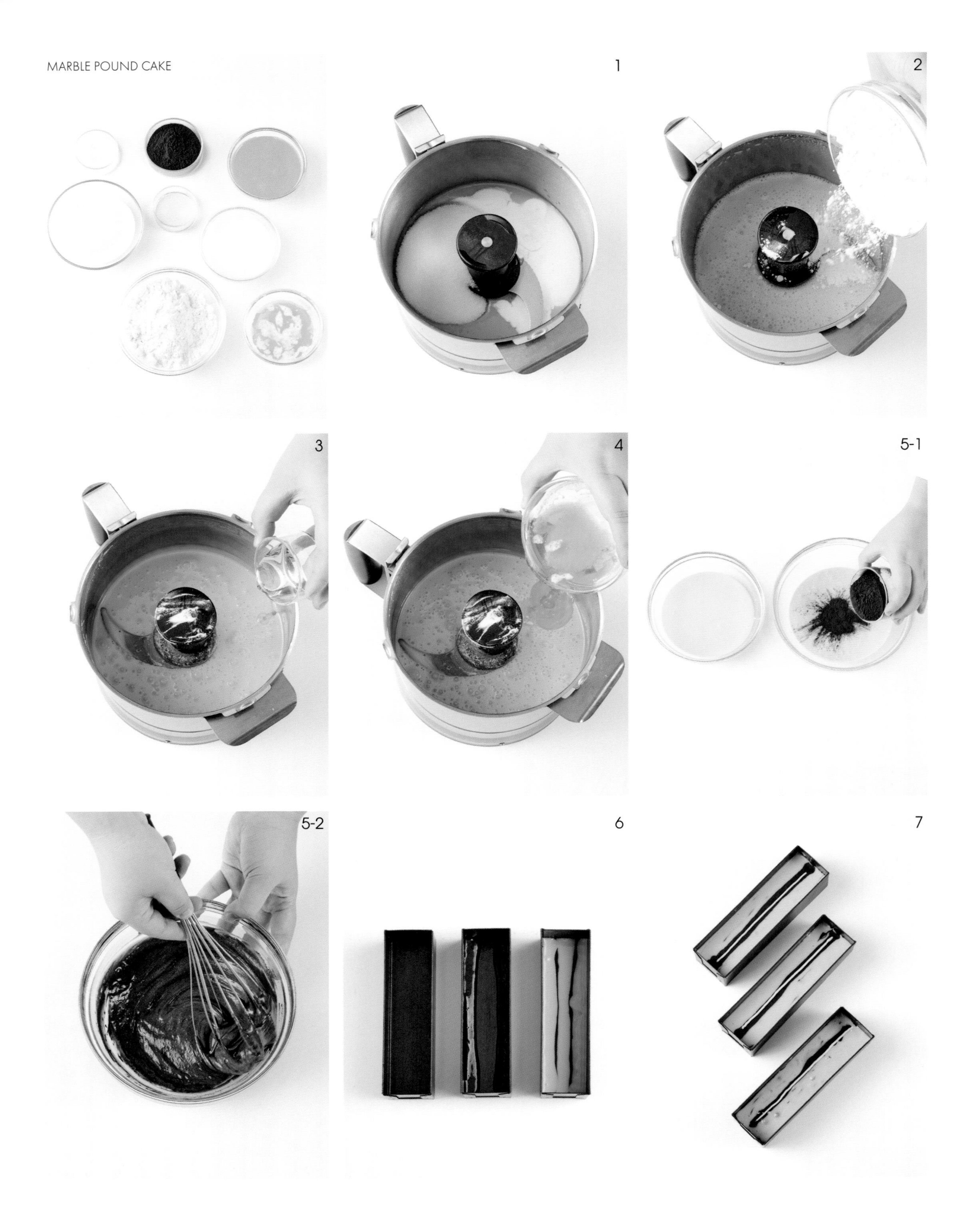
1
2
3
4
5-1
5-2
6
7

Process

마블 파운드 케이크

1. 푸드프로세서에 달걀노른자, 설탕, 플레인요거트를 넣고 믹싱한다.
2. 체 친 박력분, 베이킹파우더를 넣고 믹싱한다.
3. 포도씨유를 넣고 믹싱한다.
4. 녹인 버터(55℃)를 넣고 믹싱한다.
5. 반죽을 반으로 나눈 후 하나의 반죽에 카카오파우더를 넣고 혼합한다.
6. 버터를 얇게 칠한 틀(16.5 × 4.5 × 4.5cm)에 두 가지 반죽을 교차하며 팬닝한다.
   (파운드 케이크 1개 – 총 190g)
7. 반죽 중심부에 실온 상태의 버터(분량 외)를 가늘게 파이핑한 후 165℃ 오븐에서 약 25분간 굽는다.

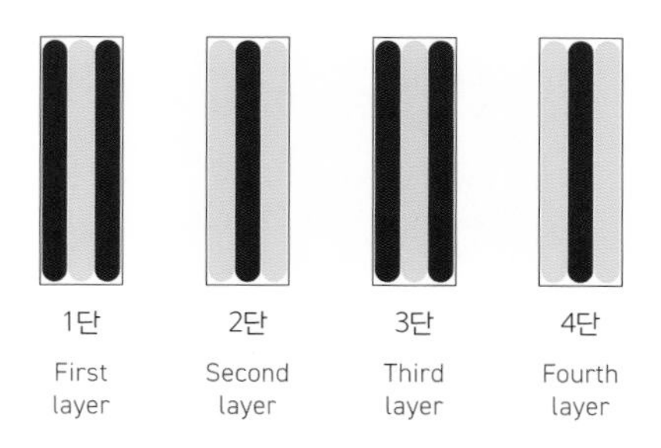

MARBLE POUND CAKE

1. Mix egg yolks, sugar, and plain yogurt in a food processor.
2. Add sifted flour and baking powder.
3. Pour in grapeseed oil and mix.
4. Add melted butter (55°C) to mix.
5. Divide the batter into two and combine cacao powder with one of the batter.
6. Pipe the two batters into lightly greased molds (16.5 × 4.5 × 4.5 cm) alternatively.
   (Total of 190g per cake)
7. Pipe a thin line of room temperature butter in the center (other than requested) and bake for about 25 minutes at 165°C.

GOLD RUM SYRUP
8
9
BITTER CHOCOLATE GANACHE
10
11
12
13
BRAUN
14

골드럼 시럽

8. 모든 재료를 혼합한다.
9. 구워져 나온 파운드 케이크에 시럽을 듬뿍 바른 후 완전히 식힌다.

비터 초콜릿 가나슈

10. 냄비에 생크림, 전화당을 넣고 60℃ 정도로 가열한다.
11. 녹인 다크초콜릿(35℃)에 **10**을 나눠 부으면서 혼합한다.
12. 실온 상태의 버터를 넣고 핸드블렌더로 혼합한다.
13. 럼을 넣고 핸드블렌더로 혼합한다.
14. 트레이에 균일한 높이로 부어준 후 밀착 랩핑한다. 파이핑하기 적합한 상태가 될 때까지 냉장고에 보관한 후 사용한다.

GOLD RUM SYRUP

8. Combine all the ingredients.
9. Soak the baked cakes generously with the syrup and let cool completely.

BITTER CHOCOLATE GANACHE

10. Heat heavy cream and inverted sugar in a saucepan to about 60°C.
11. Gradually stir in the melted dark chocolate (35°C).
12. Add room-temperature butter and blend with an immersion blender.
13. Add rum and combine with the blender.
14. Pour onto a tray in even thickness and cover with plastic wrap, making sure it adheres to the surface. Refrigerate until the texture is consistent enough to use in a piping bag.

15
16
17
18
19
20
21
22

헤이즐넛 프랄리네

15. 헤이즐넛과 바닐라빈은 140℃ 오븐에서 약 8분간 구워준다.
16. 냄비에 설탕, 물을 넣고 가열한다.
17. 118~121℃가 되면 따뜻한 상태의 헤이즐넛을 넣고 섞어준다.
18. 시럽이 하얗게 재결정화 상태가 될 때까지 계속해서 저어준다.
19. 다시 불에 올려 골고루 저어주며 진한 갈색이 날 때까지 계속해서 가열한다.
20. 실팻에 넓게 펼쳐 완전히 식힌다.
21. 푸드프로세서에 넣고 게랑드소금과 함께 갈아준다.
22. 흐르는 정도의 페이스트 상태가 되면 마무리한다.

HAZELNUT PRALINE

15. Bake hazelnuts and vanilla beans for about 8 minutes at 140°C.
16. Heat sugar and water in a saucepan.
17. When it becomes 118~121°C, remove from the heat and stir in the hot hazelnuts.
18. Continue to stir until the sugar recrystallizes and turns white.
19. Put it back over the heat and stir thoroughly until it becomes dark brown.
20. Spread onto a Silpat and cool completely.
21. Blend in a food processor with Guérande salt.
22. Finish when the paste obtains flowing consistency.

몰딩

23. 준비한 몰드에 템퍼링한 다크 초콜릿을 가득 채운 후 얇은 막이 형성될 때까지 기다린다.

    * 몰드는 사용하기 약 15분 전에 냉동고에 넣어두어 차가운 상태로 준비한다.

24. 몰드를 뒤집어 여분의 초콜릿을 털어낸다.
25. 스크래퍼를 이용해 주입구를 깔끔하게 정리한다.
26. 초콜릿이 완전히 굳을 때까지 그대로 둔다.

25

26

MOLDING

23. Completely fill with tempered dark chocolate in the prepared mold and wait until a thin layer forms.

    * Keep the mold in a freezer for 15 minutes before use to keep cold.

24. Turn over to shake off the excess chocolate.
25. Use a scraper to clean the rim of the mold.
26. Let stand until hardened.

ASSEMBLY

32

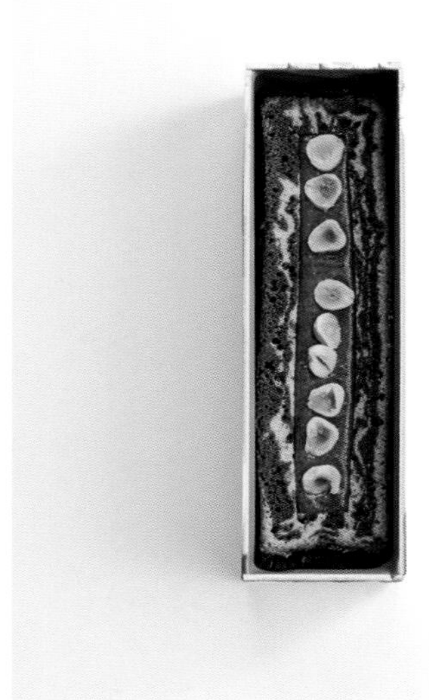

33

34

조립

27. 몰딩한 초콜릿이 완전히 굳으면 비터 초콜릿 가나슈를 균일하게 파이핑한다.
28. 1.5cm 두께로 자른 파운드 케이크의 가장 상단을 가나슈 위에 얹고 공기가 차지 않도록 눌러가며 가나슈에 접착시킨다.
29. 비터 초콜릿 가나슈를 파이핑한다.
30. 1.5cm 두께로 자른 후 정중앙에 1cm 정도로 홈을 낸 파운드 케이크를 얹고 공기가 차지 않도록 눌러가며 가나슈에 접착시킨다.
31. 홈에 헤이즐넛 프랄리네를 가득 채워준다.
32. 헤이즐넛 프랄리네 위에 구운 헤이즐넛을 가지런히 올려준다.
33. 비터 초콜릿 가나슈를 파이핑한다.
34. 1.5cm 두께로 자른 파운드 케이크 하단을 가나슈 위에 얹고 공기가 차지 않도록 눌러가며 가나슈에 접착시킨 후 냉장고에서 약 30분간 굳힌다.

ASSEMBLY

27. When the chocolate is completely hardened, evenly pipe bitter chocolate ganache.
28. Place the very top part of the pound cake, sliced into 1.5 cm thickness, over the ganache and press to adhere while making sure air does not remain inside.
29. Pipe bitter chocolate ganache.
30. Cut the pound cake to 1.5 cm thickness and make a hole in the center about 1 cm wide. Place it over the ganache and press to adhere while making sure air does not remain inside.
31. Fill the hole with hazelnut praliné.
32. Arrange roasted hazelnuts over the hazelnut praliné.
33. Pipe bitter chocolate ganache.
34. Place the bottom part of the pound cake, sliced into 1.5 cm thickness, over the ganache and press to adhere while making sure air does not remain inside. Refrigerate for about 30 minutes.

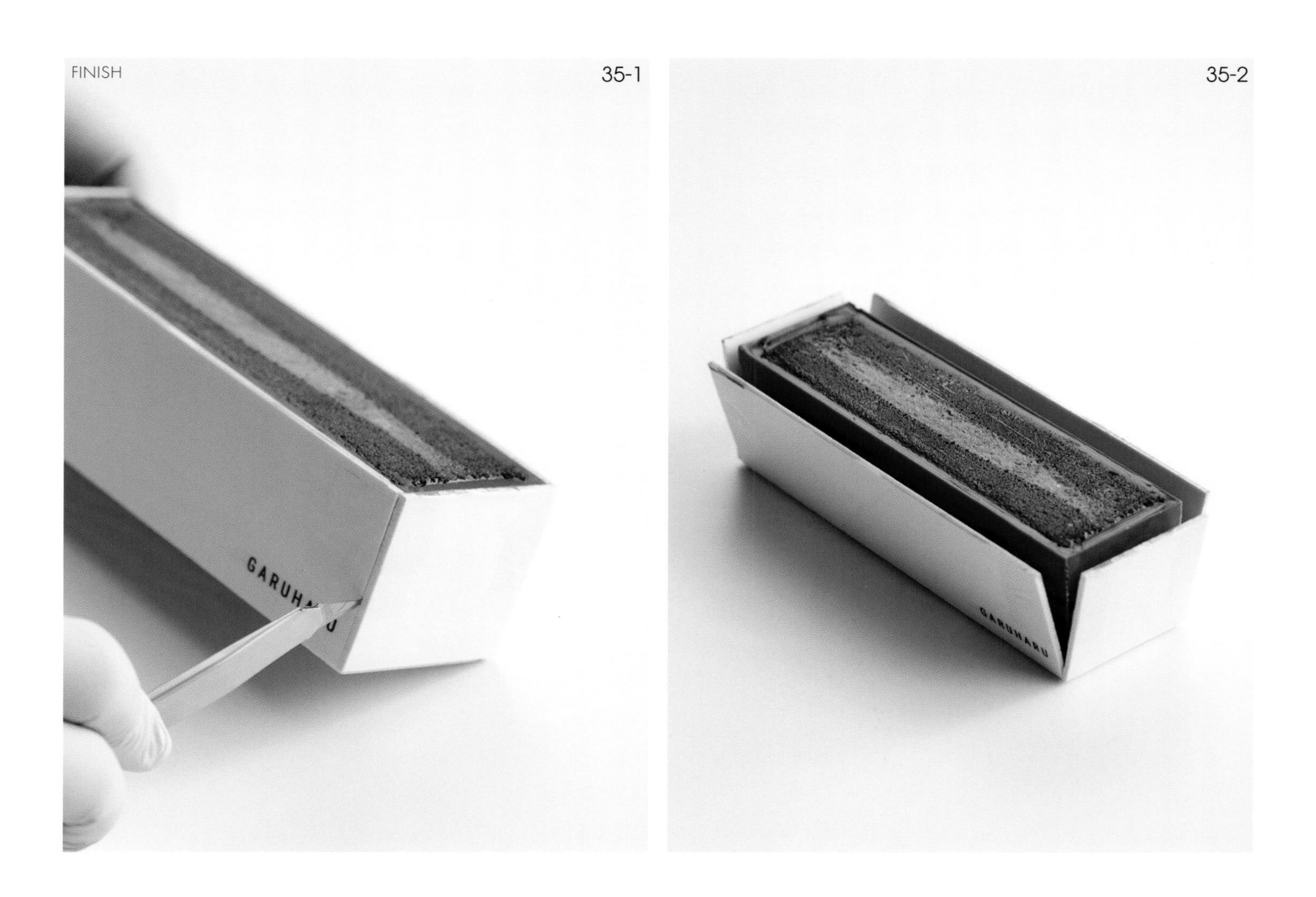

마무리

35. 몰드와 파운드 케이크를 분리한다.

36. 거친 솔로 파운드 케이크 윗면에 스크래치를 낸다.

37. 브러시로 초콜릿 가루를 털어낸 후 열풍기로 가볍게 열을 가해 표면을 매끄럽게 정리한다.

FINISH

35. Separate the cake from the mold.
36. Scratch the top of the cake with a wire brush.
37. Brush off the excess chocolate powder and lightly apply heat with a heat gun to organize the surface.

2 MATCHA POUND CAKE

말차 파운드 케이크

ingredients - 16.5 × 4.5 × 4.5cm, 3 cakes

말차 파운드 케이크

버터 139g
설탕A 105g
달걀노른자 49g
달걀전란 29g
플레인요거트 43g
달걀흰자 78g
설탕B 29g

아몬드파우더 98g
박력분 98g
말차파우더 21g
(클로렐라 함유)
베이킹파우더 1.4g
오렌지 리큐어 4g
(COINTREAU)

오렌지 리큐어 시럽

30보메 시럽 200g
물 30g
오렌지 리큐어 115g
(COINTREAU)

기타

통팥 앙금
화이트초콜릿
말차파우더
(클로렐라 함유)

말차 가나슈

화이트초콜릿 235g
(OPALYS 33%)
카카오버터 10g
생크림 175g
전화당 13g
말차파우더 10g
(클로렐라 함유)
버터 50g
오렌지 리큐어 10g
(COINTREAU)

MATCHA POUND CAKE

139g Butter
105g Sugar A
49g Egg yolks
29g Whole eggs
43g Plain yogurt
78g Egg whites
29g Sugar B

98g Almond powder
98g Cake flour
21g Matcha powder
(with chlorella)
1.4g Baking powder
4g Orange liqueur
(COINTREAU)

ORANGE LIQUEUR SYRUP

200g 30°B syrup
30g Water
115g Orange liqueur
(COINTREAU)

OTHER

Sweet whole red beans
White chocolate
Matcha powder
(with chlorella)

MATCHA GANACHE

235g White chocolate
(OPALYS 33%)
10g Cacao butter
175g Heavy cream
13g Inverted sugar
10g Matcha powder
(with chlorella)
50g Butter
10g Orange liqueur
(COINTREAU)

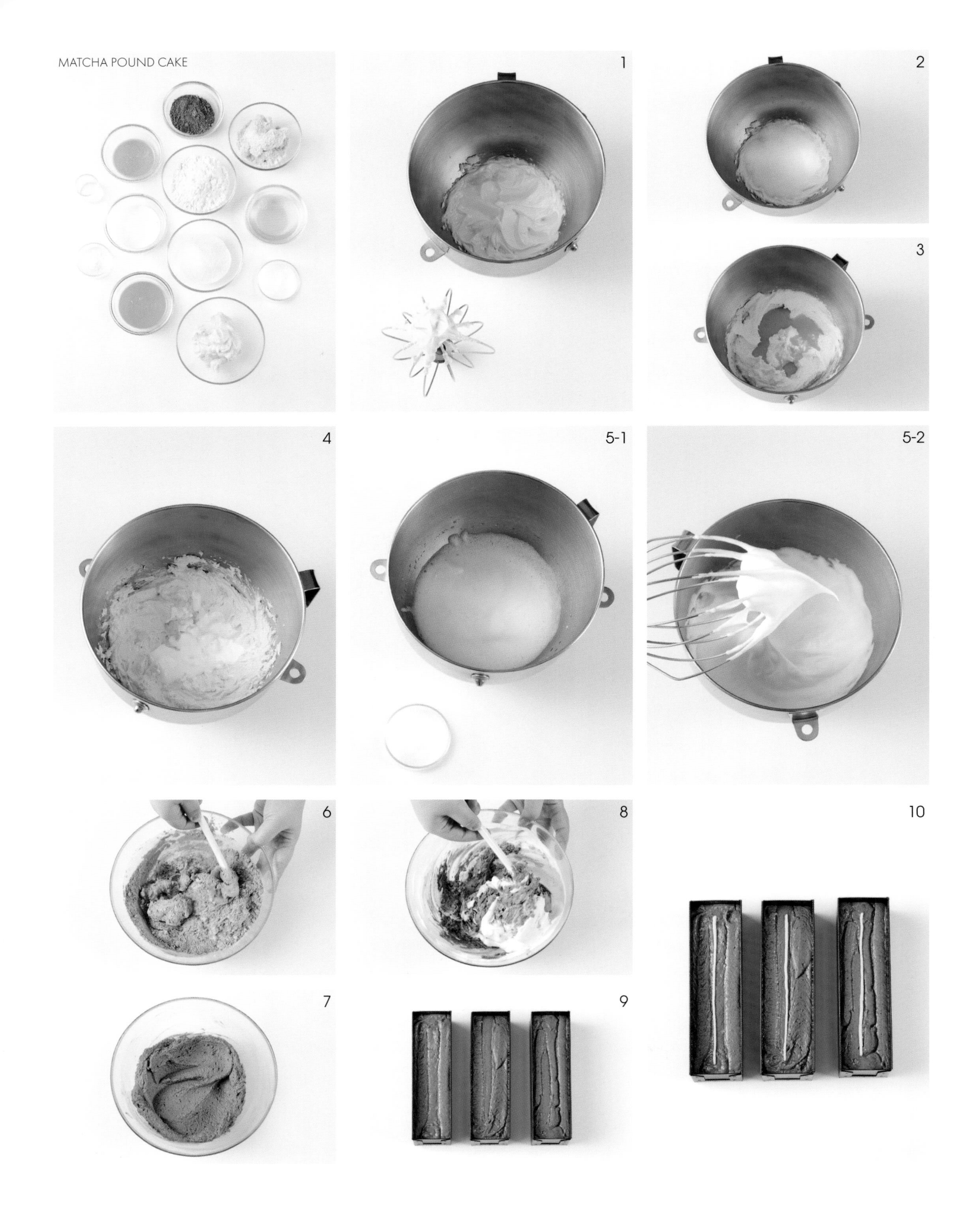
MATCHA POUND CAKE
1
2
3
4
5-1
5-2
6
8
10
7
9

Process

말차 파운드 케이크

1. 실온 상태의 버터를 부드럽게 풀어준다.
2. 설탕A를 넣고 믹싱한다.
3. 달걀노른자 – 달걀전란 순서로 조금씩 나눠 넣어가며 믹싱한다.
4. 플레인요거트를 넣고 믹싱한다.
5. 다른 볼에 달걀흰자를 넣고 거품 상태까지 믹싱한 후 설탕B를 조금씩 나눠 넣어가며 힘 있는 머랭을 완성한다.
6. **4**에 체 친 아몬드파우더, 박력분, 말차파우더, 베이킹파우더를 넣고 섞어준다.
7. 오렌지 리큐어를 넣고 섞어준다.
8. 머랭을 두 번에 나눠 넣어가며 혼합한다.
9. 버터를 얇게 칠한 틀(16.5 × 4.5 × 4.5cm)에 완성된 반죽을 210g씩 팬닝한다.
10. 반죽 중심부에 실온 상태의 버터(분량 외)를 가늘게 파이핑한 후 165℃ 오븐에서 약 25분간 굽는다.

MATCHA POUND CAKE

1. Soften room temperature butter.
2. Mix with sugar A.
3. Gradually mix in order of egg yolks – whole eggs, a little bit at a time.
4. Mix with plain yogurt.
5. In a separate bowl, whip egg whites until frothy and add sugar B, little by little, to make meringue with stiff peaks.
6. Combine sifted almond powder, cake flour, matcha powder, and baking powder into **4**.
7. Add orange liqueur.
8. Add the meringue, half at a time.
9. Pour 210 grams of finished batter into the lightly greased molds (16.5 × 4.5 × 4.5 cm).
10. Pipe a thin line of room temperature butter in the center (other than requested) and bake for about 25 minutes at 165°C.

ORANGE LIQUEUR SYRUP
11
12
MATCHA GANACHE
13
14
15
16
17
18
19

오렌지 리큐어 시럽

11. 모든 재료를 혼합한다.
12. 구워져 나온 파운드 케이크에 시럽을 듬뿍 바른 후 완전히 식힌다.

말차 가나슈

13. 녹인 화이트초콜릿(35℃)에 녹인 카카오버터(35℃)를 섞어준다.
14. 냄비에 생크림, 전화당을 넣고 60℃ 정도로 가열한다.
15. 말차파우더를 넣고 잘 섞어준 후 체에 내린다.
16. **13**에 조금씩 나눠 넣어가며 혼합한다.
17. 실온 상태의 버터를 넣고 핸드블렌더로 혼합한다.
18. 오렌지 리큐어를 넣고 계속해서 핸드블렌더로 혼합한다.
19. 트레이에 균일한 높이로 부어준 후 밀착 랩핑한다. 파이핑하기 적합한 상태가 될 때까지 냉장고에 보관한 후 사용한다.

ORANGE LIQUEUR SYRUP

11. Mix all of the ingredients.
12. Soak the baked cakes generously with the syrup and let cool completely.

MATCHA GANACHE

13. Mix melted cacao butter (35°C) into melted white chocolate (35°C).
14. Heat heavy cream and inverted sugar to about 60°C.
15. Mix with matcha powder thoroughly and filter through a sieve.
16. Gradually add to **13**, a little bit at a time.
17. Add room-temperature butter and combine with an immersion blender.
18. Add orange liqueur and continue to blend.
19. Pour onto a tray in even thickness and cover with plastic wrap, making sure it adheres to the surface. Refrigerate until the texture is consistent enough to use in a piping bag.

MOLDING
20
21
ASSEMBLY
22
23
24
25
26
27
28

몰딩

20. 템퍼링한 화이트초콜릿에 말차파우더를 넣고 핸드블렌더로 균일하게 혼합한다.
21. 52-53p와 동일한 방법으로 준비한 몰드에 몰딩 작업을 한다.

조립

22. 몰딩한 초콜릿이 완전히 굳으면 말차 가나슈를 균일하게 파이핑한다.
23. 1.5cm 두께로 자른 파운드 케이크의 가장 상단을 가나슈 위에 얹고 공기가 차지 않도록 눌러가며 가나슈에 접착시킨다.
24. 말차 가나슈를 파이핑한다.
25. 1.5cm 두께로 자른 후 정중앙에 1cm 정도로 홈을 낸 파운드 케이크를 얹고 공기가 차지 않도록 눌러가며 가나슈에 접착시킨다.
26. 홈에 통팥 앙금을 가득 채워준다.
27. 말차 가나슈를 파이핑한다.
28. 1.5cm 두께로 자른 파운드 케이크 하단을 가나슈 위에 얹고 공기가 차지 않도록 눌러가며 가나슈에 접착시킨 후 냉장고에서 약 30분간 굳힌다.

MOLDING

20. Add matcha powder to the tempered white chocolate and homogenize with an immersion blender.
21. Coat the prepared mold, as on page 52-53.

ASSEMBLY

22. When the chocolate is completely hardened, evenly pipe matcha ganache.
23. Place the very top part of the pound cake, sliced into 1.5 cm thickness, over the ganache and press to adhere while making sure air does not remain inside.
24. Pipe matcha ganache.
25. Cut the pound cake to 1.5 cm thickness and make a hole in the center about 1 cm wide. Place it over the ganache and press to adhere.
26. Fill the hole with sweet whole red beans.
27. Pipe matcha ganache.
28. Place the bottom part of the pound cake, sliced into 1.5 cm thickness, over the ganache and press to adhere while making sure air does not remain inside. Refrigerate for about 30 minutes.

마무리

29. 몰드와 파운드 케이크를 분리한 후 거친 솔로 파운드 케이크 윗면에 스크래치를 낸다.
30. 브러시로 초콜릿 가루를 털어낸 후 열풍기로 가볍게 열을 가해 표면을 매끄럽게 정리한다.
31. 냉장고에 약 10분간 두었다가 꺼낸 후 파운드 케이크 윗면에 말차파우더를 부린다.
32. 여분의 말차파우더는 브러시를 이용해 정리한다.

FINISH

29. Separate the cake from the mold, and scratch the top of the cake with a wire brush.
30. Brush off the excess chocolate powder and lightly apply heat with a heat gun to organize the surface.
31. Refrigerate for about 10 minutes and dust the top with matcha powder.
32. Brush off excess matcha powder.

3 MONT BLANC POUND CAKE

몽블랑 파운드 케이크

ingredients - 16.5 × 4.5 × 4.5cm, 3 cakes

몽블랑 파운드 케이크

버터 109g
원당 50g
무스코바도 설탕 64g
밤 페이스트 68g
밤 퓌레 68g
아몬드파우더 64g
달걀노른자 43g
달걀전란 55g
박력분 116g
베이킹파우더 3.5g
다크럼 13.7g
(NEGRITA)
보늬밤 137g

다크럼 시럽

30보메 시럽 200g
물 30g
다크럼 115g
(NEGRITA)

기타

화이트초콜릿
(OPALYS 33%)
바닐라빈 페이스트
(NOROHY VANILLA BEAN PASTE - MADAGASCAR)

밤 가나슈

생크림 77g
전화당 13.7g
밀크초콜릿 86g
(JIVARA LATTE 40%)
밤 페이스트 257g
버터 51g
커피 익스트랙 10g
다크럼 5g
(NEGRITA)

MONT BLANC POUND CAKE

109g Butter
50g Raw sugar
64g Muscovado sugar
68g Chestnut paste
68g Chestnut puree
64g Almond powder
43g Egg yolks
55g Whole eggs
116g Cake Flour
3.5g Baking powder
13.7g Dark rum
(NEGRITA)
137g Chestnuts in syrup

DARK RUM SYRUP

200g 30°B syrup
30g Water
115g Dark rum
(NEGRITA)

OTHER

White chocolate
(OPALYS 33%)
Vanilla bean paste
(NOROHY VANILLA BEAN PASTE - MADAGASCAR)

CHESTNUT GANACHE

77g Heavy cream
13.7g Inverted sugar
86g Milk chocolate
(JIVARA LATTE 40%)
257g Chestnut paste
51g Butter
10g Coffee extract
5g Dark rum
(NEGRITA)

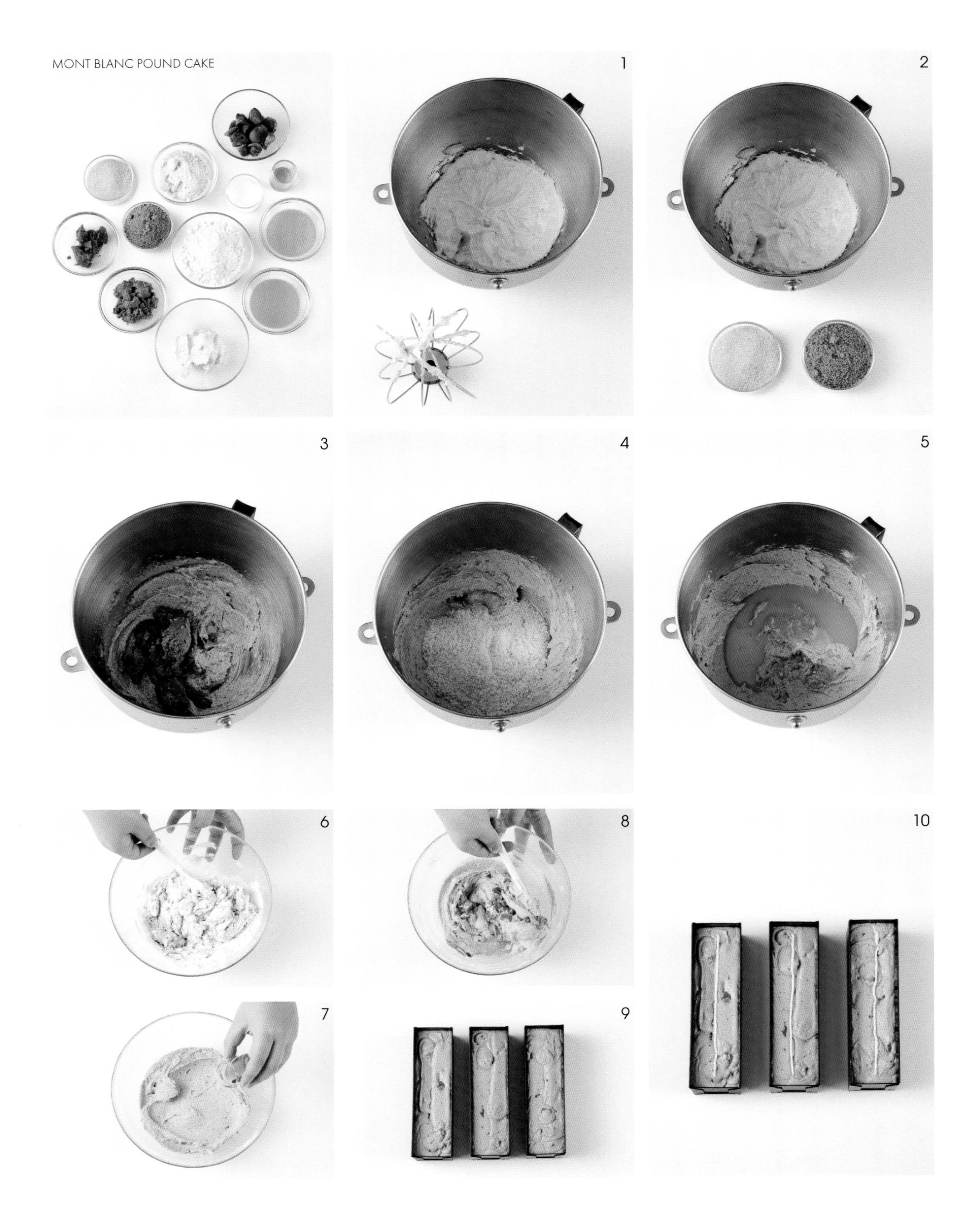
1
2
3
4
5
6
8
10
7
9

Process

몽블랑 파운드 케이크

1. 실온 상태의 버터를 부드럽게 풀어준다.
2. 원당, 무스코바도 설탕을 넣고 믹싱한다.
3. 밤 페이스트, 밤 퓌레를 넣고 믹싱한다.
4. 체 친 아몬드파우더를 넣고 믹싱한다.
5. 달걀노른자 – 달걀전란 순서로 나눠 넣어가며 믹싱한다.
6. 체 친 박력분, 베이킹파우더를 넣고 섞어준다.
7. 다크럼을 넣고 섞어준다.
8. 사방 0.5cm 크기로 다진 보늬밤을 넣고 섞어준다.
9. 버터를 얇게 칠한 틀(16.5 × 4.5 × 4.5cm)에 완성된 반죽을 240g씩 팬닝한다.
10. 반죽 중심부에 실온 상태의 버터(분량 외)를 가늘게 파이핑한 후 165℃ 오븐에서 약 25분간 굽는다.

MONT BLANC POUND CAKE

1. Soften room temperature butter.
2. Add raw sugar and muscovado sugar.
3. Add chestnut paste and chestnut puree.
4. Add sifted almond powder.
5. Gradually mix in order of egg yolks - whole eggs, a little bit a time.
6. Sift and add cake flour and baking powder.
7. Add dark rum to mix.
8. Add and combine the chestnuts cut into 0.5 cm cubes.
9. Pour 240 grams of finished batter into the lightly greased molds (16.5 × 4.5 × 4.5 cm).
10. Pipe a thin line of room temperature butter in the center (other than requested) and bake for about 25 minutes at 165°C.

DARK RUM SYRUP
11
12
CHESTNUT GANACHE
13
14
15
16
17
18

다크 럼 시럽

11. 모든 재료를 혼합한다.
12. 구워져 나온 파운드 케이크에 시럽을 듬뿍 바른 후 완전히 식힌다.

밤 가나슈

13. 냄비에 생크림, 전화당을 넣고 60℃ 정도로 가열한다.
14. 푸드프로세서에 녹인 밀크초콜릿(35℃), 밤 페이스트를 넣고 **13**을 조금씩 나눠 넣어가며 혼합한다.
15. 실온 상태의 버터를 넣고 혼합한다.
16. 커피 익스트랙을 넣고 혼합한다.
17. 다크럼을 넣고 혼합한다.
18. 트레이에 균일한 높이로 부어준 후 밀착 랩핑한다. 파이핑하기 적합한 상태가 될 때까지 냉장고에 보관한 후 사용한다.

DARK RUM SYRUP

11. Combine all the ingredients.
12. Soak the baked cakes generously with the syrup and let cool completely.

CHESTNUT GANACHE

13. Heat heavy cream and inverted sugar to about 60°C.
14. In a food processor, pour milk chocolate melted to 35°C and chestnut paste; combine while gradually adding **13** a little bit at a time.
15. Add room-temperature butter and combine.
16. Add coffee extract and continue to blend.
17. Add dark rum to mix.
18. Pour onto a tray in even thickness and cover with plastic wrap, making sure it adheres to the surface. Refrigerate until the texture is consistent enough to use in a piping bag.

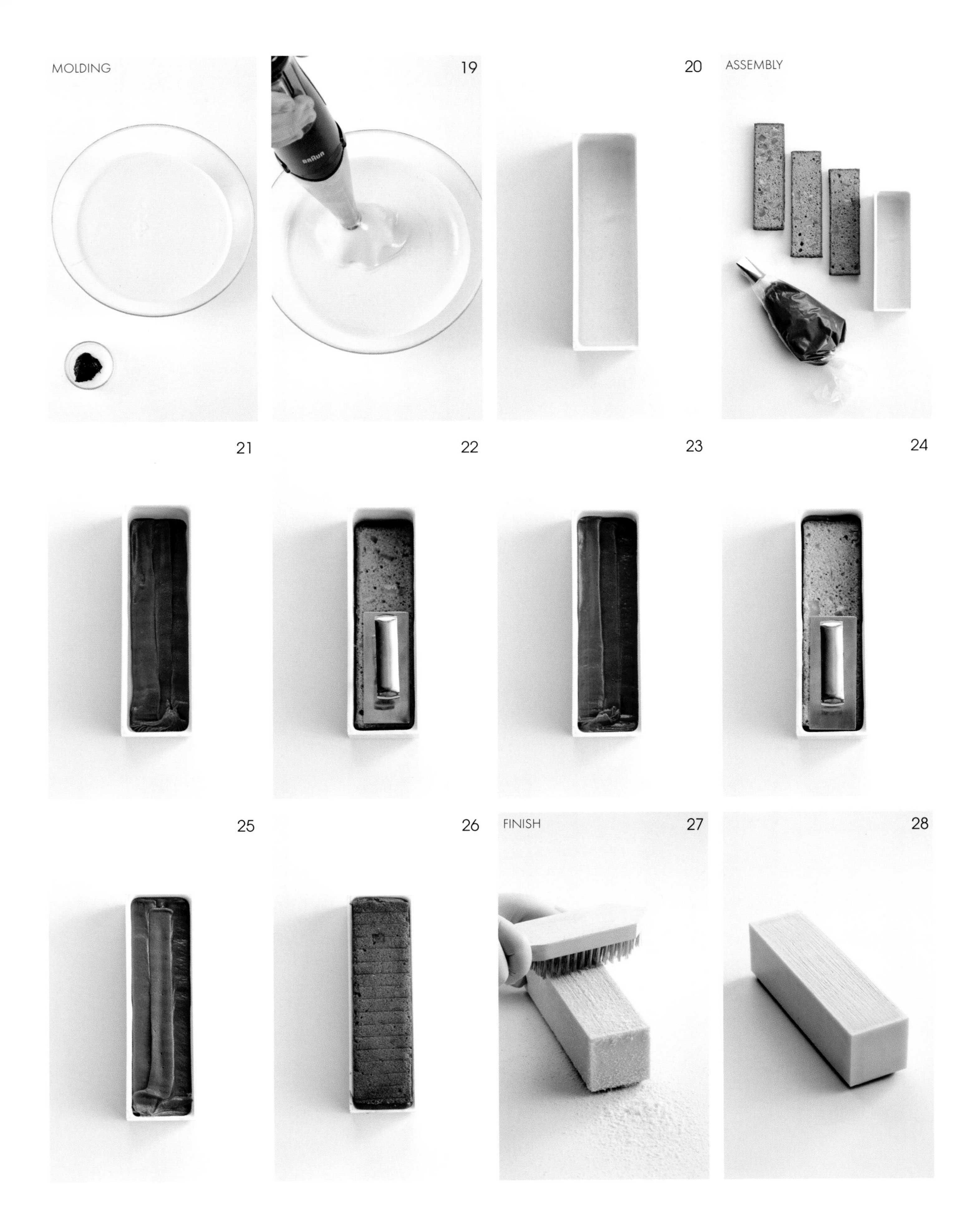
MOLDING
19
20
ASSEMBLY
21
22
23
24
25
26
FINISH
27
28

몰딩

19. 템퍼링한 화이트초콜릿에 바닐라빈 페이스트를 넣고 핸드블렌더를 이용해 균일하게 혼합한다.
20. 52-53p와 동일한 방법으로 몰드에 몰딩 작업을 한다.

조립

21. 몰딩한 초콜릿이 완전히 굳으면 밤 가나슈를 균일하게 파이핑한다.
22. 1.5cm 두께로 자른 파운드 케이크의 가장 상단을 가나슈 위에 얹고 공기가 차지 않도록 눌러가며 가나슈에 접착시킨다.
23. 밤 가나슈를 파이핑한다.
24. 1.5cm 두께로 자른 파운드 케이크의 중간 단을 가나슈 위에 얹고 공기가 차지 않도록 눌러가며 가나슈에 접착시킨다.
25. 밤 가나슈를 파이핑한다.
26. 1.5cm 두께로 자른 파운드 케이크 하단을 가나슈 위에 얹고 공기가 차지 않도록 눌러가며 가나슈에 접착시킨 후 냉장고에서 약 30분간 굳힌다.

마무리

27. 몰드와 파운드 케이크를 분리한 후 거친 솔로 파운드 케이크 윗면에 스크래치를 낸다.
28. 브러시로 초콜릿 가루를 털어낸 후 열풍기로 가볍게 열을 가해 표면을 매끄럽게 정리한다.

MOLDING

19. Add vanilla bean paste to the tempered white chocolate and combine with an immersion blender.
20. Coat the prepared mold, as on page 52-53.

ASSEMBLY

21. When the chocolate is completely hardened, evenly pipe chestnut ganache.
22. Place the very top part of the pound cake, sliced into 1.5 cm thickness, over the ganache and press to adhere while making sure air does not remain inside.
23. Pipe chestnut ganache.
24. Cut the middle section of the pound cake to 1.5 cm thickness, place it over the ganache, and press to adhere while making sure air does not remain inside.
25. Pipe chestnut ganache.
26. Place the bottom part of the pound cake, sliced into 1.5 cm thickness, over the ganache and press to adhere while making sure air does not remain inside. Refrigerate for about 30 minutes.

FINISH

27. Separate the cake from the mold and scratch the top of the cake with a wire brush.
28. Brush off the excess chocolate powder and lightly apply heat with a heat gun to organize the surface.

GARUHARU

4 YUJA POUND CAKE

유자 파운드 케이크

ingredients - 16.5 × 4.5 × 4.5cm, 3 cakes

유자 파운드 케이크

버터 126g
설탕 126g
소금 2.1g
아몬드파우더 44g
플레인요거트 31g
달걀노른자 47g
달걀전란 47g
박력분 175g
베이킹파우더 2.6g
유자주스 27g

유자 시럽

30보메 시럽 200g
유자주스 30g
패션프루트 리큐어 115g
(DIJON FRUIT DE LA PASSION)

유자 가나슈

생크림 107g
전화당 25g
유자초콜릿 245g
(YUZU INSPIRATION)
카카오버터 31g
유자주스 46g
버터 46g

기타

유자초콜릿
(YUZU INSPIRATION)
노란색 천연 식용 색소

YUJA POUND CAKE

126g Butter
126g Sugar
2.1g Salt
44g Almond powder
31g Plain yogurt
47g Egg yolks
47g Whole eggs
175g Cake flour
2.6g Baking powder
27g Yuja juice

YUJA SYRUP

200g 30°B syrup
30g Yuja juice
115g Passion fruit liqueur
(DIJON FRUIT DE LA PASSION)

YUJA GANACHE

107g Heavy cream
25g Inverted sugar
245g Yuzu chocolate
(YUZU INSPIRATION)
31g Cacao butter
46g Yuja juice
46g Butter

OTHER

Yuzu chocolate
(YUZU INSPIRATION)
Yellow natural food color

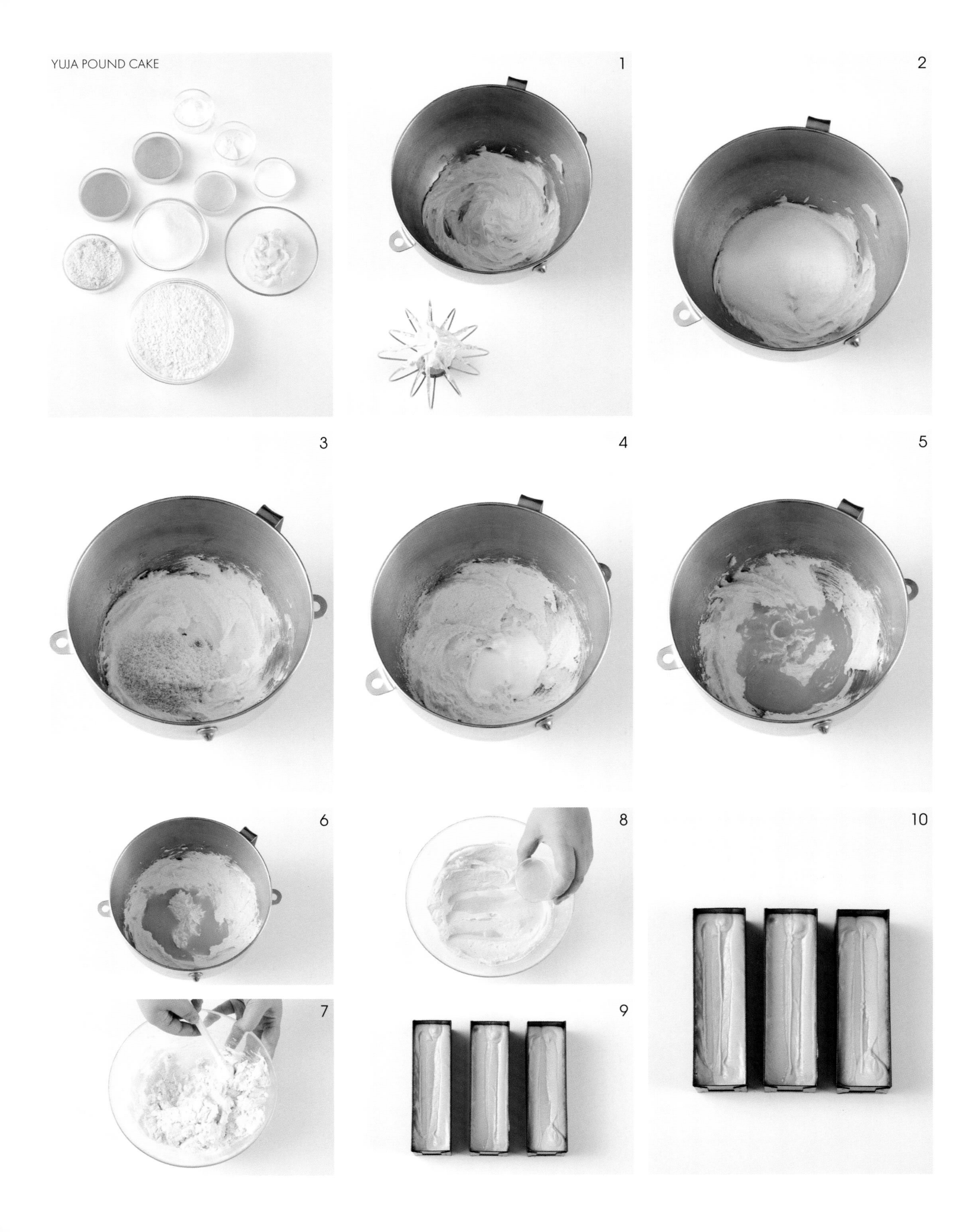
YUJA POUND CAKE
1
2
3
4
5
6
7
8
9
10

Process

유자 파운드 케이크

1. 실온 상태의 버터를 부드럽게 풀어준다.
2. 설탕, 소금을 넣고 믹싱한다.
3. 체 친 아몬드파우더를 넣고 믹싱한다.
4. 플레인요거트를 넣고 믹싱한다.
5. 달걀노른자를 넣고 믹싱한다.
6. 달걀전란을 나눠 넣어가며 믹싱한다.
7. 체 친 박력분, 베이킹파우더를 넣고 섞어준다.
8. 유자주스를 넣고 섞어준다.
9. 버터를 얇게 칠한 틀(16.5 × 4.5 × 4.5cm)에 완성된 반죽을 190g씩 팬닝한다.
10. 반죽 중심부에 실온 상태의 버터(분량 외)를 가늘게 파이핑한 후 165℃ 오븐에서 약 25분간 굽는다.

YUJA POUND CAKE

1. Soften room-temperature butter.
2. Mix with sugar and salt.
3. Sift and add almond powder and mix.
4. Add plain yogurt.
5. Add and mix egg yolks.
6. Gradually add whole eggs while mixing.
7. Sift and add cake flour and baking powder.
8. Combine with yuja juice.
9. Pour 190 grams of finished batter into the lightly greased molds (16.5 × 4.5 × 4.5 cm).
10. Pipe a thin line of room temperature butter in the center (other than requested) and bake for about 25 minutes at 165°C.

YUJA SYRUP
11
12
YUJA GANACHE
13
14
15
16
17
18

유자 시럽

11. 모든 재료를 혼합한다.
12. 구워져 나온 파운드 케이크에 시럽을 듬뿍 바른 후 완전히 식힌다.

유자 가나슈

13. 냄비에 생크림, 전화당을 넣고 60℃ 정도로 가열한다.
14. 볼에 녹인 유자초콜릿(35℃)과 녹인 카카오버터(35℃)를 넣고 혼합한다.
15. **14**에 **13**을 조금씩 넣어가며 섞는다.
16. 데운 유자주스(30℃)를 넣고 핸드블렌더를 이용해 혼합한다.
17. 실온 상태의 버터를 넣고 계속해서 혼합한다.
18. 트레이에 균일한 높이로 부어준 후 밀착 랩핑한다. 파이핑하기 적합한 상태가 될 때까지 냉장고에 보관한 후 사용한다.

YUJA SYRUP

11. Combine all the ingredients.
12. Soak the baked cakes generously with the syrup and let cool completely.

YUJA GANACHE

13. Heat heavy cream and inverted sugar to about 60°C.
14. Combine melted Yuzu chocolate (35°C) and cacao butter (35°C).
15. Add **13** into **14**, a little bit at a time.
16. Add warmed yuja juice (30°C) and blend using an immersion blender.
17. Add room-temperature butter and continue to blend.
18. Pour onto a tray in even thickness and cover with plastic wrap, making sure it adheres to the surface. Refrigerate until the texture is consistent enough to use in a piping bag.

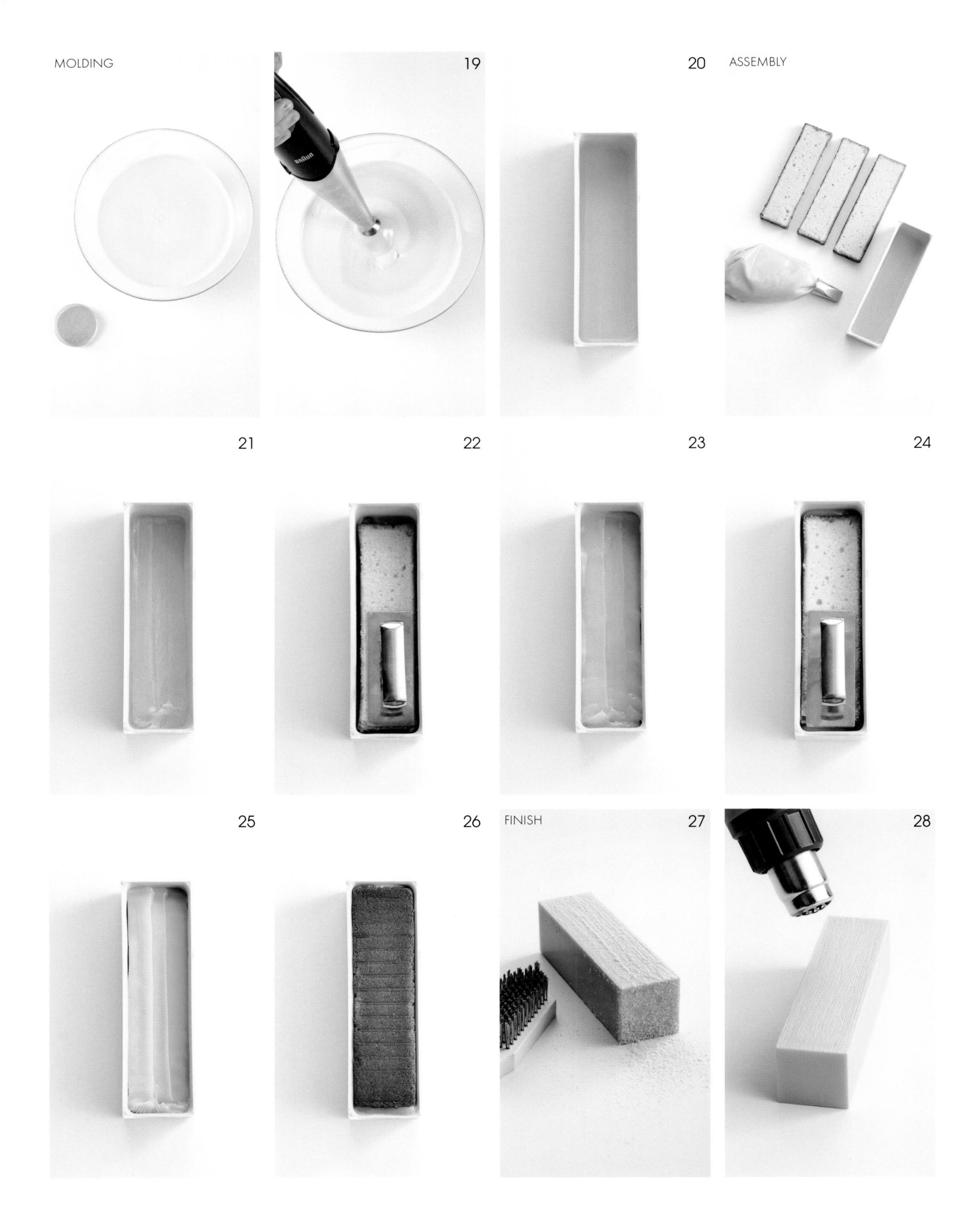

MOLDING
19
20
ASSEMBLY
21
22
23
24
25
26
FINISH
27
28

몰딩

19. 템퍼링한 유자초콜릿에 노란색 천연 식용 색소를 넣고 핸드블렌더를 이용해 균일하게 혼합한다.
20. 52-53p와 동일한 방법으로 몰드에 몰딩 작업을 한다.

조립

21. 몰딩한 초콜릿이 완전히 굳으면 유자 가나슈를 균일하게 파이핑한다.
22. 1.5cm 두께로 자른 파운드 케이크의 가장 상단을 가나슈 위에 얹고 공기가 차지 않도록 눌러가며 가나슈에 접착시킨다.
23. 유자 가나슈를 파이핑한다.
24. 1.5cm 두께로 자른 파운드 케이크의 중간 단을 가나슈 위에 얹고 공기가 차지 않도록 눌러가며 가나슈에 접착시킨다.
25. 유자 가나슈를 파이핑한다.
26. 1.5cm 두께로 자른 파운드 케이크 하단을 가나슈 위에 얹고 공기가 차지 않도록 눌러가며 가나슈에 접착시킨 후 냉장고에서 약 30분간 굳힌다.

마무리

27. 몰드와 파운드 케이크를 분리한 후 거친 솔로 파운드 케이크 윗면에 스크래치를 낸다.
28. 브러시로 초콜릿 가루를 털어낸 후 열풍기로 가볍게 열을 가해 표면을 매끄럽게 정리한다.

MOLDING

19. Add yellow natural food color to the tempered Yuzu chocolate and combine with an immersion blender.
20. Coat the prepared mold, as on page 52-53.

ASSEMBLY

21. When the chocolate is completely hardened, evenly pipe yuja ganache.
22. Place the very top part of the pound cake, sliced into 1.5 cm thickness, over the ganache and press to adhere while making sure air does not remain inside.
23. Pipe yuja ganache.
24. Cut the middle section of the pound cake to 1.5 cm thickness, place it over the ganache, and press to adhere while making sure air does not remain inside.
25. Pipe yuja ganache.
26. Place the bottom part of the pound cake, sliced into 1.5 cm thickness, over the ganache and press to adhere while making sure air does not remain inside. Refrigerate for about 30 minutes.

FINISH

27. Separate the cake from the mold and scratch the top of the cake with a wire brush.
28. Brush off the excess chocolate powder and lightly apply heat with a heat gun to organize the surface.

5 FIG POUND CAKE

무화과 파운드 케이크

ingredients - 16.5 × 4.5 × 4.5cm, 3 cakes

NUTS FREE

무화과 페이스트*

반건조 무화과 267g
레드와인 233g
(CABERNET SAUVIGNON)

무화과 파운드 케이크

버터 141g
슈거파우더 134g
달걀전란 96g
박력분 120g
카카오파우더 32g
베이킹파우더 3.6g
무화과 페이스트* 265g

양배 시럽

30보메 시럽 200g
배주스 30g
양배 리큐어 115g
(DIJON POIRES WILLIAM)

기타

프람보아즈초콜릿
(FRAMBOISE INSPIRATION)

무화과 충전물

설탕 58g
트레할로스 40g
NH펙틴 8.3g
라즈베리 퓌레 166g
무화과 퓌레 90g
배주스 45g
레몬주스 61g
꿀 31g
무화과 페이스트* 140g

NUTS FREE

FIG PASTE*

267g Semi-dried figs
233g Red wine
(CABERNET SAUVIGNON)

FIG POUND CAKE

141g Butter
134g Powdered sugar
96g Whole eggs
120g Cake flour
32g Cacao powder
3.6g Baking powder
265g Fig paste*

PEAR SYRUP

200g 30°B syrup
30g Pear juice
115g Pear liqueur
(DIJON POIRES WILLIAM)

OTHER

Framboise chocolate
(FRAMBOISE INSPIRATION)

FIG FILLING

58g Sugar
40g Trehalose
8.3g Pectin NH
166g Raspberry puree
90g Fig puree
45g Pear juice
61g Lemon juice
31g Honey
140g Fig paste*

1
2
3
4
5
6

Process

무화과 페이스트

1. 반건조 무화과의 꼭지를 제거한다.
2. 뾰족한 도구로 무화과에 구멍을 낸다.
3. 냄비에 무화과가 잠길 정도의 물을 넣고 가열한다.
4. 손질한 반건조 무화과를 데친 후 물기를 제거한다.
5. 푸드프로세서에 데친 무화과, 레드와인을 넣고 갈아준다.
6. 균일한 질감의 페이스트 상태로 마무리한다.

FIG PASTE

1. Remove the stems from the figs.
2. Make holes in the figs using a pointy tool.
3. Add enough water to cover the figs in a pot and heat.
4. Blanch the trimmed semi-dried figs and drain.
5. Blend the blanched figs and red wine in a food processor.
6. Finish when the paste is homogeneous.

7
8
9
10
11

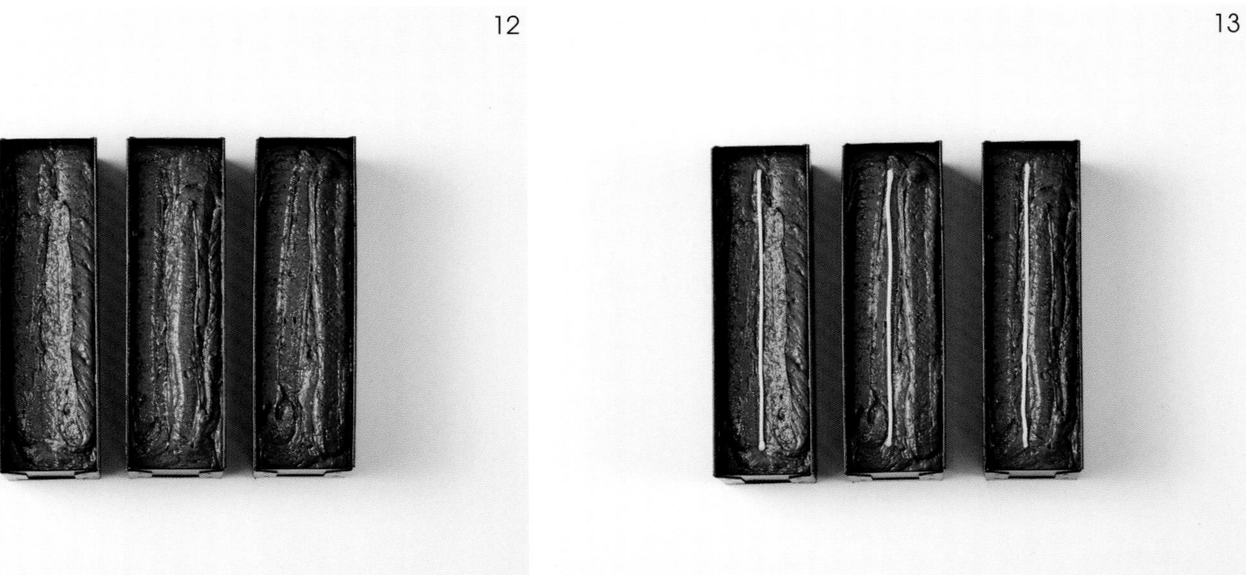
12
13

**무화과
파운드 케이크**

7. 실온 상태의 버터를 부드럽게 풀어준다.
8. 체 친 슈거파우더를 넣고 믹싱한다.
9. 달걀전란을 조금씩 나눠 넣어가며 믹싱한다.
10. 체 친 박력분, 카카오파우더, 베이킹파우더를 넣고 섞어준다.
11. 무화과 페이스트를 넣고 섞어준다.
12. 버터를 얇게 칠한 틀(16.5 × 4.5 × 4.5cm)에 완성된 반죽을 240g씩 팬닝한다.
13. 반죽 중심부에 실온 상태의 버터(분량 외)를 가늘게 파이핑한 후 165℃ 오븐에서 약 30분간 굽는다.

**FIG
POUND
CAKE**

7. Soften room temperature butter.
8. Sift and add powdered sugar.
9. Gradually add whole eggs to mix, a little bit at a time.
10. Sift and add cake flour, cacao powder, and baking powder.
11. Add and combine fig paste.
12. Pour 240 grams of finished batter into the lightly greased molds (16.5 × 4.5 × 4.5 cm).
13. Pipe a thin line of room temperature butter in the center (other than requested) and bake for about 30 minutes at 165°C.

PEAR SYRUP
14
15
FIG FILLING
16
17
18
19
20
21

양배 시럽

14. 모든 재료를 혼합한다.
15. 구워져 나온 파운드 케이크에 시럽을 듬뿍 바른 후 완전히 식힌다.

무화과 충전물

16. 설탕, 트레할로스, NH펙틴을 섞어준다.
17. 냄비에 라즈베리 퓌레, 무화과 퓌레, 배주스, 레몬주스, 꿀을 넣고 가열한다.
18. 40℃가 되면 **16**을 넣고 섞는다.
19. 계속해서 저어가며 40Brix가 될 때까지 가열한다.
20. 무화과 페이스트를 넣고 섞어준다.
21. 트레이에 부어 차갑게 식힌 후 부드럽게 풀어 사용한다.

PEAR SYRUP

14. Combine all the ingredients.
15. Soak the baked cakes generously with the syrup and let cool completely.

FIG FILLING

16. Combine sugar, trehalose, and pectin NH.
17. Heat raspberry puree, fig puree, pear juice, lemon juice, and honey in a pot.
18. When the mixture is 40°C, stir in **16**.
19. Continue to heat while stirring until it reaches 40 Brix.
20. Add fig paste to combine.
21. Pour over a tray, let cool completely, and soften to use.

MOLDING

22

ASSEMBLY

23

24

25

26

27

28

FINISH

29

30

몰딩 22. 템퍼링한 프람보아즈초콜릿을 이용해 52-53p와 동일한 방법으로 몰드에 몰딩 작업을 한다.

조립 23. 몰딩한 초콜릿이 완전히 굳으면 무화과 충전물을 균일하게 파이핑한다.

24. 1.5cm 두께로 자른 파운드 케이크의 가장 상단을 무화과 충전물 위에 얹고 공기가 차지 않도록 눌러가며 무화과 충전물에 접착시킨다.

25. 무화과 충전물을 파이핑한다.

26. 1.5cm 두께로 자른 파운드 케이크의 중간 단을 무화과 충전물 위에 얹고 공기가 차지 않도록 눌러가며 무화과 충전물에 접착시킨다.

27. 무화과 충전물을 파이핑한다.

28. 1.5cm 두께로 자른 파운드 케이크 하단을 무화과 충전물 위에 얹고 공기가 차지 않도록 눌러가며 무화과 충전물에 접착시킨 후 냉장고에서 약 30분간 굳힌다.

마무리 29. 몰드와 파운드 케이크를 분리한 후 거친 솔로 파운드 케이크 윗면에 스크래치를 낸다.

30. 브러시로 초콜릿 가루를 털어낸 후 열풍기로 가볍게 열을 가해 표면을 매끄럽게 정리한다.

MOLDING 22. Using the tempered Framboise chocolate, coat the prepared mold, as on page 52-53.

ASSEMBLY 23. When the chocolate is completely hardened, pipe fig filling evenly.

24. Place the very top part of the pound cake, sliced into 1.5 cm thickness, over the filling and press to adhere while making sure air does not remain inside.

25. Pipe fig filling.

26. Cut the middle section of the pound cake to 1.5 cm thickness, place it over the filling, and press to adhere while making sure air does not remain inside.

27. Pipe fig filling.

28. Place the bottom part of the pound cake, sliced into 1.5 cm thickness, over the filling and press to adhere while making sure air does not remain inside. Refrigerate for about 30 minutes.

FINISH 29. Separate the cake from the mold and scratch the top of the cake with a wire brush.

30. Brush off the excess chocolate powder and lightly apply heat with a heat gun to organize the surface.

GARUHARU

GARUHARU
GARUHARU
GARUHARU

6 PISTACHIO & GRAPEFRUIT CAKE

피스타치오 & 자몽 케이크

ingredients - 12 cakes

GLUTEN FREE

피스타치오 케이크

달걀흰자 247g
바닐라빈 페이스트 0.4g
(NOROHY VANILLA BEAN PASTE - MADAGASCAR)
고운 쌀가루 114g
베이킹파우더 2.6g
슈거파우더 118g
아몬드파우더 114g
(Valencia)
자몽제스트 1.1g
꿀 63g
소금 2g
피스타치오 페이스트 44g
버터 123g

자몽 시럽

30보메 시럽 200g
자몽주스 30g
오렌지 리큐어 115g
(Grand Marnier)

자몽 충전물

전처리한 자몽 166g
설탕 58g
트레할로스 40g
NH펙틴 8.3g
자몽주스 90g
유자주스 72g
레몬주스 34g
꿀 31g

초콜릿 글레이즈(실온 유지)

화이트초콜릿 319g
(OPALYS 33%)
화이트 컴파운드초콜릿 192g
포도씨유 45g
피스타치오 페이스트 45g

기타

오렌지 필
(SABATON)
피스타치오 커넬

GLUTEN FREE

PISTACHIO CAKE

247g Egg whites
0.4g Vanilla bean paste
(NOROHY VANILLA BEAN PASTE - MADAGASCAR)
114g Fine rice flour
2.6g Baking powder
118g Powdered sugar
114g Almond powder
(Valencia)
1.1g Grapefruit zest
63g Honey
2g Salt
44g Pistachio paste
123g Butter

GRAPEFRUIT SYRUP

200g 30°B syrup
30g Grapefruit juice
115g Orange liqueur
(Grand Marnier)

GRAPEFRUIT FILLING

166g Pre-treated grapefruit
58g Sugar
40g Trehalose
8.3g Pectin NH
90g Grapefruit juice
72g Yuja juice
34g Lemon juice
31g Honey

CHOCOLATE GLAZE (MAINTAIN ROOM TEMPERATURE)

319g White chocolate
(OPALYS 33%)
192g White compound chocolate
45g Grapeseed oil
45g Pistachio paste

OTHER

Candied orange peels
(SABATON)
Pistachio kernels

PISTACHIO CAKE
1
2
3
4
5
6-1
6-2
7

Process

피스타치오 케이크

1. 달걀흰자, 바닐라빈 페이스트를 핸드블렌더로 혼합한다.
2. 체 친 고운 쌀가루, 베이킹파우더, 슈거파우더, 아몬드파우더, 자몽제스트를 섞어준다.
3. 1을 조금씩 나눠 넣어가며 섞어준다.
4. 꿀을 넣고 섞어준다.
5. 소금, 피스타치오 페이스트를 넣고 섞어준다.
   * 구운 피스타치오 페이스트, 굽지 않은 피스타치오 페이스트를 절반씩 사용한다.
6. 녹인 버터(55℃)를 넣고 섞어준다.
7. 버터를 얇게 칠한 미니 구겔호프 틀에 65g씩 팬닝한 후 160℃ 오븐에서 25분간 굽는다.

PISTACHIO CAKE

1. Combine egg whites and vanilla bean paste with an immersion blender.
2. Mix with sifted fine rice flour, baking powder, powdered sugar, almond powder, and grapefruit zest.
3. Gradually add with **1**, little by little.
4. Add honey.
5. Mix with salt and pistachio paste.
   * Use half of the roasted pistachio paste and half of the raw pistachio paste.
6. Mix with melted butter (55°C).
7. Pour 65 grams into the lightly greased mini Gugelhupf molds and bake for 25 minutes at 160°C.

GRAPEFRUIT SYRUP
8
9
GRAPEFRUIT FILLING
10
11
12
13
14
15
16

자몽 시럽

8. 모든 재료를 혼합한다.
9. 구워져 나온 케이크에 시럽을 듬뿍 적시고 완전히 식힌 후 냉동고에 보관한다.

자몽 충전물

10. 자몽은 끓는 물에 넣고 껍질이 완전히 부드러워질 때까지 약 1시간 30분간 삶는다.
    * 이때 15분마다 새 물로 교체한다.
11. 체에 걸러 물기를 제거한 후 잘게 썰어 준비한다. (166g)
12. 설탕, 트레할로스, NH펙틴을 섞어준다.
13. 냄비에 **11**과 자몽주스, 유자주스, 레몬주스, 꿀을 넣고 가열한다.
14. 40℃가 되면 **12**를 넣고 혼합한 후 45Brix가 될 때까지 가열한다.
15. 핸드블렌더로 곱게 갈아준다.
16. 트레이에 부어 차갑게 식힌 후 부드럽게 풀어 사용한다.

GRAPEFRUIT SYRUP

8. Combine all the ingredients.
9. Soak the baked cakes generously with the syrup and store them in the freezer.

GRAPEFRUIT FILLING

10. Boil grapefruits in boiling water and cook until skin becomes soft, about 1 hour and 30 minutes.
    * Replace the water every 15 minutes.
11. Drain to remove water and chop finely. (166g)
12. Combine sugar, trehalose, and pectin NH in a separate bowl.
13. Heat **11**, grapefruit juice, yuja juice, lemon juice, and honey in a pot.
14. When the mixture is 40°C, combine with **12** and heat until it becomes 45 Brix.
15. Blend finely with an immersion blender.
16. Pour onto a tray, cool completely, and soften to use.

CHOCOLATE GLAZE
(MAINTAIN ROOM TEMPERATURE)
17
BRAUN
18
FINISH
19-1
19-2
20
21

초콜릿 글레이즈 (실온 유지)

17. 녹인 초콜릿(40℃), 포도씨유, 피스타치오 페이스트를 혼합한다.
18. 35℃로 맞춰 사용한다.

마무리

19. 차가운 상태의 피스타치오 케이크를 초콜릿 글레이즈(35℃)로 코팅한다.
20. 초콜릿이 완전히 굳기 전에 오렌지필과 피스타치오 커넬로 장식한다.
21. 자몽 충전물을 모양내어 파이핑한다.

CHOCOLATE GLAZE (MAINTAIN ROOM TEMPERATURE)

17. Combine melted chocolate (40°C), grapeseed oil, and pistachio paste.
18. Use at 35°C.

FINISH

19. Coat the cold pistachio cakes with the chocolate glaze (35°C).
20. Decorate with orange peels and pistachio kernels before the glaze sets.
21. Pipe the grapefruit filling to finish.

7 CARROT & ORANGE CAKE

당근 & 오렌지 케이크

ingredients - 12 cakes

NUTS FREE

당근 케이크

무스코바도 설탕 117g
오렌지제스트 5g
달걀전란 138g
소금 1g
박력분 164g
베이킹파우더 3.2g
시나몬파우더 1g
플레인 요거트 106g
포도씨유 127g
잘게 다진 당근 74g
피칸 63g

오렌지 당근 시럽

설탕 150g
오렌지주스 75g
레몬주스 38g
당근주스 38g
오렌지 리큐어 75g
(Grand Marnier)

치즈 크림

크림치즈 119g
(Quescrem)
설탕 13g
연유 21g
생크림 139g
레몬주스 9g

당근 장식물

화이트초콜릿
(OPALYS 33%)
주황색 천연 식용 색소
말차파우더

초콜릿 글레이즈 (실온 유지)

화이트초콜릿 345g
(OPALYS 33%)
화이트 컴파운드초콜릿 207g
포도씨유 48g
건조 당근 큐브 5g
(당근 주스를 짜고 남은 섬유소를 50℃ 오븐에서 약 2시간 건조해 사용)

기타

시나몬파우더

NUTS FREE

CARROT CAKE

117g Muscovado sugar
5g Orange zest
138g Whole eggs
1g Salt
164g Cake flour
3.2g Baking powder
1g Cinnamon powder
106g Plain yogurt
127g Grapeseed oil
74g Carrots, finely chopped
63g Pecans

ORANGE CARROT SYRUP

150g Sugar
75g Orange juice
38g Lemon juice
38g Carrot juice
75g Orange liqueur
(Grand Marnier)

CHEESE CREAM

119g Cream cheese
(Quescrem)
13g Sugar
21g Condensed milk
139g Heavy cream
9g Lemon juice

CARROT DECORATION

White chocolate
(OPALYS 33%)
Orange natural food color
Matcha powder

OTHER

Cinnamon powder

CHOCOLATE GLAZE (MAINTAIN ROOM TEMPERATURE)

345g White chocolate
(OPALYS 33%)
207g White compound chocolate
48g Grapeseed oil
5g Dried carrot cubes
(Dry the remaining fiber after making juice for about 2 hours at 50°C to use.)

CARROT CAKE
1
2
3
4
5
6
7
8

Process

당근 케이크

1. 피칸은 150℃ 오븐에서 약 20분간 로스팅한 후 잘게 다져 준비한다.
2. 무스코바도 설탕과 오렌지제스트를 혼합한다.
3. 믹싱볼에 **2**와 달걀전란, 소금을 넣고 충분히 믹싱한다.
4. **3**에 체 친 박력분, 베이킹파우더, 시나몬파우더를 넣고 혼합한다.
5. 플레인 요거트를 넣고 섞는다.
6. 포도씨유를 넣고 섞는다.
7. 잘게 다진 당근과 피칸을 넣고 섞는다.
8. 버터를 얇게 칠한 미니 구겔호프 틀에 완성된 반죽을 60g씩 팬닝한 후 160℃ 오븐에서 약 30분간 굽는다.

CARROT CAKE

1. Roast pecans for about 20 minutes at 150°C and chop finely.
2. Combine muscovado sugar and orange zest.
3. In a mixing bowl, thoroughly combine **2** with whole eggs and salt.
4. Add sifted cake flour, baking powder, and cinnamon powder into **3** to mix.
5. Add plain yogurt.
6. Mix with grapeseed oil.
7. Add finely chopped carrots and pecans.
8. Pour 60 grams into the lightly greased mini Gugelhupf molds and bake for about 30 minutes at 160°C.

ORANGE CARROT SYRUP
9
11
10
12
CHEESE CREAM
13
14
15
16
17

오렌지 당근 시럽

9. 냄비에 설탕과 오렌지주스를 넣고 설탕이 녹을 때까지 가열한 후 차갑게 식힌다.
10. **9**에 레몬주스, 당근주스, 오렌지 리큐어를 넣고 섞는다.
11. 구워져 나온 케이크에 시럽을 듬뿍 적신다.
12. 실온에서 완전히 식힌 후 냉동 보관한다.

치즈 크림

13. 냉장 상태의 크림치즈를 부드럽게 풀어준 후 설탕을 넣고 섞어준다.
14. 연유를 넣고 섞어준다.
15. 생크림을 조금씩 넣어가며 섞어준다.
16. 레몬주스를 넣고 섞어준다.
17. 냉장고에 보관한 후 가볍게 휘핑해 사용한다.

ORANGE CARROT SYRUP

9. Heat sugar and orange juice in a saucepan until the sugar dissolves and cool completely.
10. Add lemon juice, carrot juice, and orange liqueur to **9** and mix.
11. Soak the baked cakes generously with the syrup.
12. Let cool completely at an ambient temperature, then freeze.

CHEESE CREAM

13. Soften the cold cream cheese and combine with sugar.
14. Add condensed milk.
15. Gradually add heavy cream while mixing.
16. Add lemon juice.
17. Refrigerate and lightly whip to use.

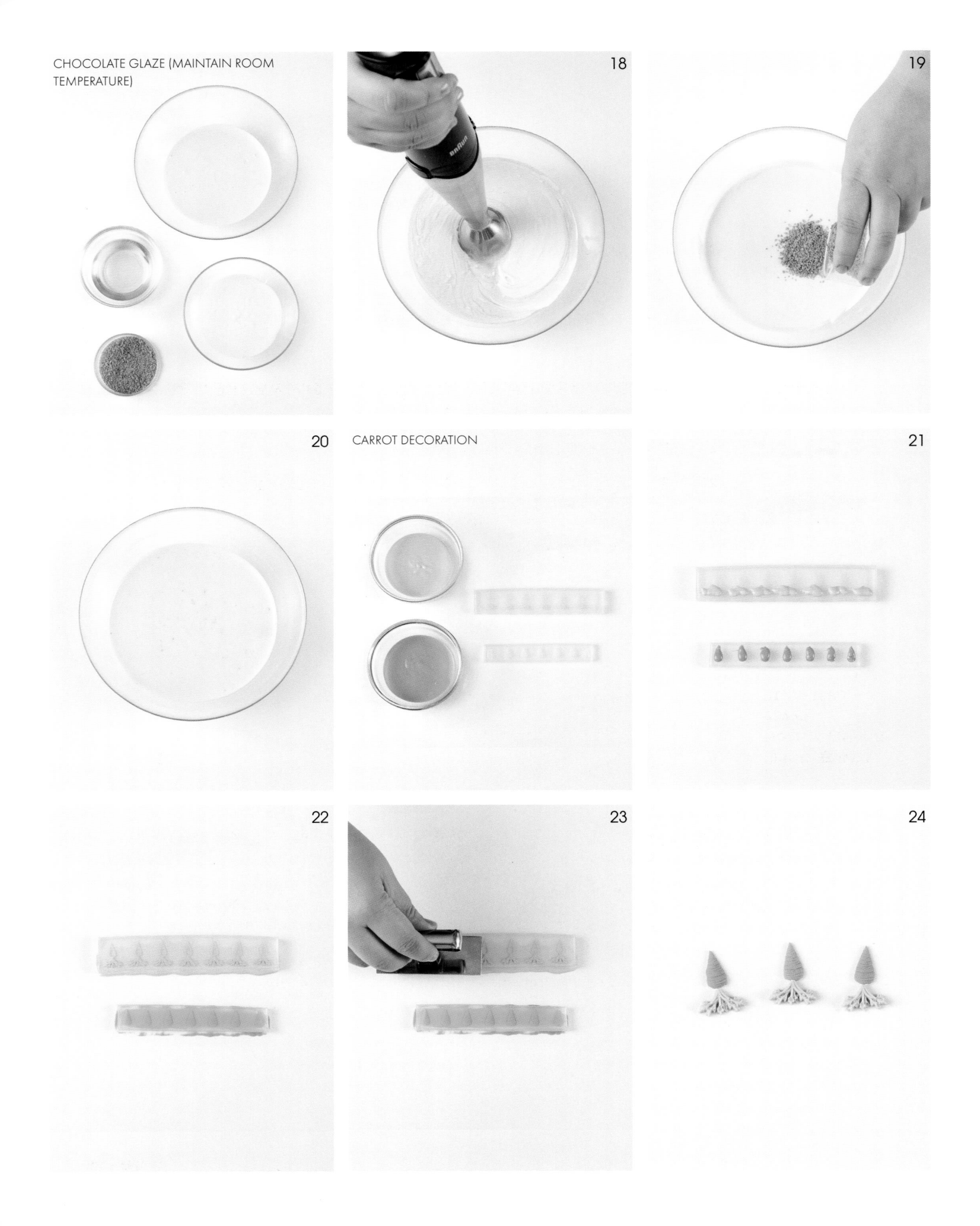
CHOCOLATE GLAZE (MAINTAIN ROOM TEMPERATURE)
18
19
20
CARROT DECORATION
21
22
23
24

초콜릿 글레이즈 (실온 유지)

18. 녹인 초콜릿(40℃), 포도씨유를 혼합한다.
19. 건조 당근 큐브를 넣고 혼합한다.
20. 35℃로 맞춰 사용한다.

당근 장식물

21. 템퍼링한 화이트초콜릿에 말차파우더, 오렌지 색소를 각각 섞어 준비한 후 당근 몸통, 잎 모양 실리콘몰드에 채운다.
22. 투명 필름 위에 몰드를 뒤집어준다.
23. 평평한 도구를 이용해 몰드를 눌러 기포를 제거하면서 몰드 안의 초콜릿이 잘 밀착되도록 한다.
24. 초콜릿이 너무 단단하게 굳기 전에 몰드를 조심스럽게 떼어낸다.

    * 완전히 굳힌 후 여분의 초콜릿을 이용해 당근 몸통과 뿌리를 접착한다.

CHOCOLATE GLAZE (MAINTAIN ROOM TEMPERATURE)

18. Combine melted chocolate (40°C) and grapeseed oil.
19. Mix with dried carrot cubes.
20. Use at 35°C.

CARROT DECORATION

21. Prepare tempered white chocolate mixed with matcha powder and orange food color and fill the carrot and leaf-shaped silicone molds.
22. Turn the molds onto the transparent film.
23. Use a flat tool to press the molds to remove any air bubbles and let the chocolate adhere to the mold.
24. Carefully remove from the molds before the chocolate sets completely.

    * Use the remaining chocolate to attach the carrots and the leaves after they are completely set.

마무리

25. 차가운 상태의 당근 케이크를 초콜릿 글레이즈(35℃)로 코팅한다.
26. 휘핑한 치즈 크림을 모양내어 파이핑한다.
27. 시나몬파우더를 뿌린다.
28. 당근 모양 초콜릿 장식물을 올린다.

FINISH

25. Coat the cold carrot cakes with the chocolate glaze (35°C).

26. Pipe the whipped cheese cream.

27. Dust with cinnamon powder.

28. Decorate with carrot-shaped chocolate decorations.

8 EARL GREY CAKE

얼그레이 케이크

ingredients - 12 cakes

얼그레이 케이크

헤이즐넛 버터 138g
달걀흰자 230g
소금 2g
슈거파우더 111g
박력분 106g
베이킹파우더 2.5g
아몬드파우더 106g
얼그레이파우더 6g
꿀 90g

얼그레이 시럽

물 100g
얼그레이 잎 3g
30보메 시럽 200g
오렌지 리큐어 115g
(Grand Marnier)

초콜릿 글레이즈 (실온 유지)

블론드초콜릿 345g
(DULCEY 35%)
화이트 컴파운드초콜릿 207g
포도씨유 48g

솔티드 버터 캐러멜 충전물

생크림 224g
게랑드소금 2.4g
바닐라빈 1/2개
(마다가스카르산)
물엿 147g
설탕 147g
젤라틴매스 7.1g
버터 73g

기타

바닐라파우더
콘 플라워

캐러멜라이즈 헤이즐넛

물 19g
설탕 64g
헤이즐넛 200g
카카오버터 8g

초콜릿 장식물

카카오파우더
블론드초콜릿
(DULCEY 35%)

EARL GREY CAKE

138g Hazelnut butter
230g Egg whites
2g Salt
111g Powdered sugar
106g Cake flour
2.5g Baking powder
106g Almond powder
6g Earl Grey tea powder
90g Honey

EARL GREY SYRUP

100g Water
3g Earl Grey tea leaves
200g 30°B syrup
115g Orange liqueur
(Grand Marnier)

CHOCOLATE GLAZE (MAINTAIN ROOM TEMPERATURE)

345g Blonde Chocolate
(DULCEY 35%)
207g White compound chocolate
48g Grapeseed oil

SALTED BUTTER CARAMEL FILLING

224g Heavy cream
2.4g Guérande salt
1/2 Vanilla bean
(Madagascar)
147g Corn syrup
147g Sugar
7.1g Gelatin mass
73g Butter

OTHER

Vanilla powder
Cornflower

CARAMELIZED HAZELNUTS

19g Water
64g Sugar
200g Hazelnuts
8g Cacao butter

CHOCOLATE DECORATION

Cacao powder
Blonde Chocolate
(DULCEY 35%)

EARL GREY CAKE
1
2
3
4
5
6
7
8

Process

얼그레이 케이크

1. 냄비에 173g의 버터를 넣고 가열한다.
2. 붉은 갈색의 헤이즐넛 버터가 완성되면 138g을 계량한 후 55℃로 맞춰 사용한다.
3. 달걀흰자, 소금을 핸드블렌더로 믹싱해 균일한 상태로 만든다.
4. 체 친 슈거파우더, 박력분, 베이킹파우더, 아몬드파우더, 얼그레이파우더를 섞어준다.
5. **4**에 **3**을 넣고 혼합한다.
6. 꿀을 넣고 섞어준다.
7. **2**를 조금씩 나눠 넣어가며 섞어준다.
8. 버터를 얇게 칠한 미니 구겔호프 틀에 완성된 반죽을 60g씩 팬닝한 후 160℃ 오븐에서 약 25분간 굽는다.

EARL GREY CAKE

1. Heat 173 grams of butter in a pot.
2. When it becomes reddish-brown hazelnut butter, weigh 138 grams, and use at 55°C.
3. Homogenize egg whites and salt using an immersion blender.
4. Sift powdered sugar, cake flour, baking powder, almond powder, and earl grey powder.
5. Combine **3** into **4**.
6. Add honey to mix.
7. Gradually add **2** to mix, little by little.
8. Pour 60 grams into the lightly greased mini Gugelhupf molds and bake for 25 minutes at 160°C.

EARL GREY SYRUP
9
10
SALTED BUTTER CARAMEL FILLING
11
12
13
14
15
17
16
18

얼그레이 시럽

9. 물과 얼그레이 잎을 수비드 백에 담아 60℃에서 1시간 동안 수비드 방식으로 인퓨징한다.
10. 체에 걸러 30g 계량한 후 30보메 시럽, 오렌지 리큐어와 섞어준다.

솔티드 버터 캐러멜 충전물

11. 냄비에 생크림, 게랑드소금, 바닐라빈을 넣고 끓기 직전까지 가열한다.
12. 다른 냄비에 물엿을 넣고 설탕을 조금씩 나눠 넣어가며 캐러멜화시킨다.
13. **11**을 조금씩 나눠 넣어가며 디글레이즈한다.
14. 108℃까지 가열한다.
15. 체에 거른다.
16. 80℃까지 식힌 후 젤라틴매스를 넣고 섞어준다.
17. 차가운 상태의 버터를 넣고 핸드블렌더로 혼합한다.
18. 실온에서 12시간 휴지시킨 후 사용한다.

EARL GREY SYRUP

9. Add water and Earl Grey tea leaves in a sous-vide bag and infuse for one hour at 60°C.
10. Filter through a sieve, weigh 30 grams and mix with 30°B syrup and orange liqueur.

SALTED BUTTER CARAMEL FILLING

11. Heat heavy cream, Guérande salt, and vanilla bean in a saucepan until just before it starts to boil.
12. In a separate pot, heat corn syrup and gradually add sugar little by little to caramelize.
13. Deglaze by adding **11**, a little bit at a time.
14. Cook to 108°C.
15. Filter through a sieve.
16. Cool to 80°C and add gelatin mass to mix.
17. Add cold butter and combine with an immersion blender.
18. Let rest for 12 hours at an ambient temperature.

19
20
21
22
23
24
25
siliko

캐러멜라이즈 헤이즐넛

19. 헤이즐넛은 140℃ 오븐에서 약 8분간 구워준다.
20. 냄비에 물과 설탕을 넣고 118~121℃까지 가열한다.
21. 따듯한 상태의 헤이즐넛을 넣고 섞어준다.
22. 시럽이 하얗게 재결정화 상태가 될 때까지 계속해서 저어준다.
23. 다시 불에 올려 골고루 저어주며 계속해서 가열한다.
24. 진한 캐러멜 색이 나면 불에서 내린 후 카카오버터를 넣고 고르게 섞어준다.
25. 실팻에 부어 펼친 후 완전히 식기 전에 한 알씩 떼어낸다.

CARAMELIZED HAZELNUTS

19. Roast the hazelnuts in an oven at 140°C for about 8 minutes.
20. In a saucepan, heat water and sugar to 118~121°C.
21. Stir in the warm hazelnuts.
22. Continue to stir until the syrup recrystallizes and turns white.
23. Put it back over the heat, and continue to stir.
24. When the color becomes a dark caramel color, remove from the heat, and mix well with cacao butter.
25. Pour over a Silpat and separate the nuts one by one before it's completely cooled.

CHOCOLATE GLAZE (MAINTAIN ROOM TEMPERATURE)
26
27
CHOCOLATE DECORATION
28
29
30

초콜릿 글레이즈 (실온 유지)

26. 녹인 초콜릿(40℃), 포도씨유를 혼합한다.
27. 35℃로 맞춰 사용한다.

초콜릿 장식물

28. 두 장의 투명 필름 사이에 카카오파우더를 뿌린다.
29. 템퍼링한 커버추어 초콜릿을 적당량 부어준다.
30. 필름을 덮고 밀대를 이용해 균일한 두께로 밀어 편 후 완전히 굳으면 자연스럽게 부서진 모양을 살려 사용한다.

CHOCOLATE GLAZE (MAINTAIN ROOM TEMPERATURE)

26. Combine melted chocolate (40°C) and grapeseed oil.
27. Use at 35°C.

CHOCOLATE DECORATION

28. Sprinkle cacao powder between two sheets of transparent films.
29. Pour a moderate amount of tempered couverture chocolate.
30. Cover with a film and roll out to even thickness. When the chocolate sets, break it naturally and preserve the uneven shapes to use.

마무리

31. 차가운 상태의 얼그레이 케이크를 초콜릿 글레이즈(35℃)로 코팅한다.

32. 바닐라파우더를 뿌린다.

33. 초콜릿이 완전히 굳기 전에 캐러멜라이즈 헤이즐넛과 콘 플라워를 붙여 고정시킨다.

34. 솔티드 버터 캐러멜 충전물을 모양내어 파이핑한 후 초콜릿 장식물로 마무리한다.

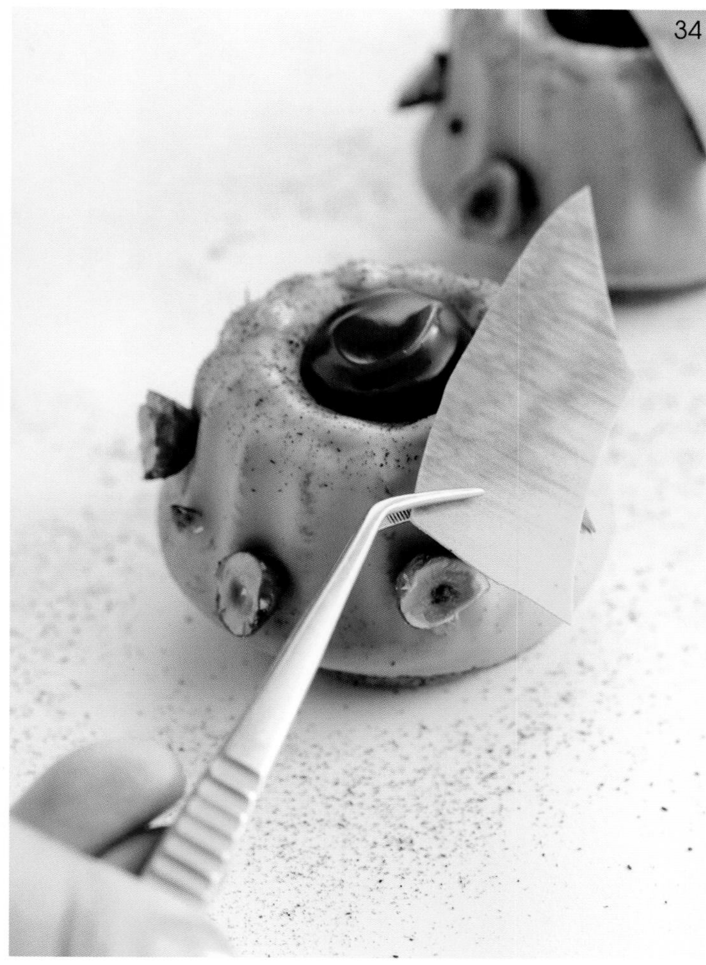

FINISH

31. Coat the cold Earl Grey cakes with the chocolate glaze (35°C).
32. Dust with vanilla powder.
33. Arrange the caramelized hazelnuts and cornflowers before the glaze sets.
34. Pipe the salted butter caramel filling and decorate with chocolate decoration to finish.

9 CACAO CAKE

카카오 케이크

ingredients - 12 cakes

카카오 케이크

버터 87g
우유 46g
달걀전란 175g
달걀노른자 69g
꿀 23g
아몬드파우더 116g
슈거파우더 180g
박력분 74g
베이킹파우더 0.5g
카카오파우더 21g

헤이즐넛 카카오 프랄리네

헤이즐넛 250g
물 58g
설탕 250g
카카오닙 25g
게랑드소금 2.5g

알콜 스프레이

다크럼
(MONARCH)

비터 초콜릿 가나슈

생크림 199g
전화당 45g
다크초콜릿 199g
(CARAIBE 66%)
버터 45g
다크럼 12g
(MONARCH)

초콜릿 글레이즈 (실온 유지)

다크초콜릿 348g
(EQUATORIALE NOIRE 55%)
다크 컴파운드초콜릿 204g
포도씨유 49g

기타

카카오닙

CACAO CAKE

87g Butter
46g Milk
175g Whole eggs
69g Egg yolks
23g Honey
116g Almond powder
180g Powdered sugar
74g Cake flour
0.5g Baking powder
21g Cacao powder

HAZELNUT CACAO PRALINE

250g Hazelnuts
58g Water
250g Sugar
25g Cacao nibs
2.5g Guérande salt

ALCOHOL SPRAY

Dark Rum
(MONARCH)

BITTER CHOCOLATE GANACHE

199g Heavy cream
45g Inverted sugar
199g Dark chocolate
(CARAIBE 66%)
45g Butter
12g Dark rum
(MONARCH)

CHOCOLATE GLAZE (MAINTAIN ROOM TEMPERATURE)

348g Dark chocolate
(EQUATORIALE NOIRE 55%)
204g Dark compound chocolate
49g Grapeseed oil

OTHER

Cacao nibs

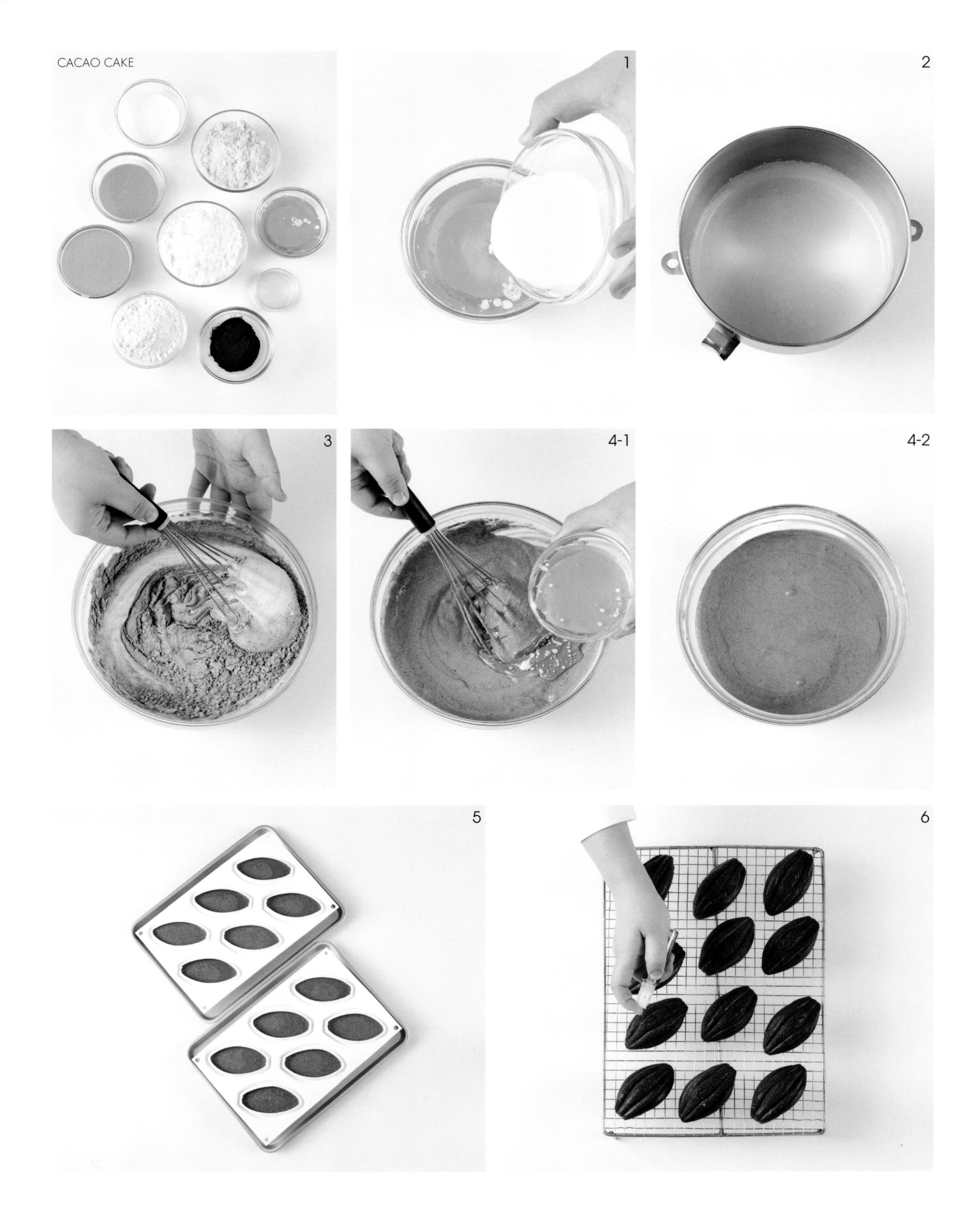
CACAO CAKE
1
2
3
4-1
4-2
5
6

Process

카카오 케이크

1. 녹인 버터(55℃)와 데운 우유(60℃)를 혼합한다.
2. 볼에 달걀전란, 달걀노른자, 꿀, 아몬드파우더, 슈거파우더를 넣고 충분히 믹싱한다.
3. 체 친 박력분, 베이킹파우더, 카카오파우더를 넣고 섞어준다.
4. 1을 조금씩 넣어가며 섞어준다.
5. 카카오 모양 실리콘 몰드(Silikomart Cacao 120)에 60g씩 팬닝한 후 160℃ 오븐에서 25분간 굽는다.
6. 구워져 나온 카카오 케이크에 알콜 스프레이를 분사한 후 실온에서 완전히 식혀 냉동고에 보관한다.

CACAO CAKE

1. Combine melted butter (55°C) and warmed milk (60°C).
2. In a bowl, add whole eggs, egg yolks, honey, almond powder, and powdered sugar; mix thoroughly.
3. Add sifted cake flour, baking powder, and cacao powder.
4. Gradually add 1 while mixing, little by little.
5. Pour 60 grams into the cacao-shaped silicone molds (Silikomart Cacao 120) and bake for 25 minutes at 160°C.
6. Apply alcohol spray on the baked cacao cakes, let cool completely at an ambient temperature, and freeze them.

7
8
9
10
11
12
13
14
15

헤이즐넛 카카오 프랄리네

7. 헤이즐넛은 140℃ 오븐에서 8분간 구워준다.
8. 냄비에 물, 설탕을 넣고 118~121℃로 가열한다.
9. 불에서 내린 후 따뜻한 상태의 **7**을 넣고 밝은 갈색이 될 때까지 섞어준다.
10. 시럽이 하얗게 재결정화 상태가 될 때까지 계속해서 저어준다.
11. 다시 불에 올려 골고루 저어주며 계속해서 열을 가한다.
12. 캐러멜화가 시작되면 카카오닙을 넣고 진한 갈색이 될 때까지 조금 더 진행한다.
13. 실팻에 넓게 펼쳐 완전히 식힌다.
14. 적당한 크기로 부순 후 게랑드소금과 함께 푸드프로세서에 넣고 갈아준다. 흐르는 정도의 페이스트 상태가 되면 마무리한다.
15. 지름 3cm 반구형 몰드에 채운 후 냉동고에서 굳힌다.

HAZELNUT CACAO PRALINE

7. Bake hazelnuts for 8 minutes at 140°C.
8. Heat water and sugar in a pot to 118~121°C.
9. Remove from heat, add warm hazelnuts (**7**), and stir until it turns golden brown.
10. Continue to stir until the syrup recrystallizes and turns white.
11. Place back over the heat and stir continuously while heating.
12. When it starts to caramelize, add cacao nibs and continue to heat until it turns dark brown.
13. Spread over a Silpat and cool completely.
14. Crush the nuts into the moderate size and grind them with Guérande salt in a food processor. Finish when it becomes a fluid paste.
15. Fill in the 3 cm diameter hemisphere mold and freeze.

BITTER CHOCOLATE GANACHE
16
17
18
19
20
CHOCOLATE GLAZE
(MAINTAIN ROOM TEMPERATURE)
21
22

비터 초콜릿 가나슈

16. 냄비에 생크림, 전화당을 넣고 60℃ 정도로 가열한다.
17. 녹인 다크초콜릿(35℃)에 **16**을 나눠 넣어가며 혼합한다.
18. 실온 상태의 버터를 넣고 핸드블렌더로 혼합한다.
19. 다크럼을 넣고 계속해서 혼합한다.
20. 트레이에 균일한 높이로 부어준 후 밀착 랩핑한다. 파이핑하기 적합한 상태가 될 때까지 냉장고에 보관한 후 사용한다.

초콜릿 글레이즈 (실온 유지)

21. 녹인 초콜릿(40℃), 포도씨유를 혼합한다.
22. 35℃로 맞춰 사용한다.

BITTER CHOCOLATE GANACHE

16. Heat heavy cream and inverted sugar in a pot to about 60°C.
17. Gradually add **16** into melted dark chocolate (35°C) to combine.
18. Add room-temperature butter and mix with an immersion blender.
19. Add dark rum and continue to mix.
20. Pour onto a tray in even thickness and cover with plastic wrap, making sure it adheres to the surface. Refrigerate until the texture is consistent enough to use in a piping bag.

CHOCOLATE GLAZE (MAINTAIN ROOM TEMPERATURE)

21. Combine melted chocolate (40°C) and grapeseed oil.
22. Use at 35°C.

23
24
25
26
27
28
29
30
31-1
31-2

마무리

23. 케이크 바닥면에 충전물을 주입할 공간을 만들어준다.
24. 비터 초콜릿 가나슈를 소량 파이핑한다.
25. 냉동고에서 굳힌 헤이즐넛 카카오 프랄리네를 넣고 가나슈를 밀착시킨다.
26. 비터 초콜릿 가나슈를 가득 채운 후 스패출러를 이용해 주입구를 매끄럽게 정리한다.
27. 케이크가 차가워지도록 냉동고에 최소 30분간 보관한다.
28. 초콜릿 글레이즈로 케이크 바닥면을 코팅한다.
29. 바닥면의 초콜릿 글레이즈가 안정적으로 굳을 때까지 냉동고에 잠시 둔다.
30. 초콜릿 글레이즈로 카카오 케이크 전체를 코팅한다.
31. 초콜릿 글레이즈가 완전히 굳기 전에 카카오닙을 뿌려 장식한다.

FINISH

23. Make a space through the bottom of the cake to inject the filling.
24. Pipe a small quantity of bitter chocolate ganache.
25. Insert the frozen hazelnut cacao praliné and attach it to the ganache.
26. Completely fill with bitter chocolate ganache and organize the opening evenly using a spatula.
27. Keep in the freezer for at least 30 minutes to chill the cakes.
28. Coat the bottom of the cake with the chocolate glaze.
29. Freeze for a while until the chocolate glaze on the bottom is stable.
30. Coat the whole cake with the chocolate glaze.
31. Sprinkle cacao nibs to decorate before the chocolate glaze sets completely.

10 COFFEE BEAN CAKE

커피 빈 케이크

ingredients - 12 cakes

커피 빈 케이크

버터 57g
에스프레소 19g
달걀전란 115g
달걀노른자 46g
꿀 16g
헤이즐넛파우더 77g
슈거파우더 119g
박력분 75g
베이킹파우더 0.3g
원두커피분말 4g

알콜 스프레이

커피 리큐어 200g
(KAHLUA)
골드럼 100g
(MONARCH)

커피 가나슈

커피원두 15g
생크림 193g
전화당 44g
다크초콜릿 193g
(CARAIBE 66%)
버터 44g
커피 리큐어 12g
(KAHLUA)

초콜릿 글레이즈

밀크초콜릿 348g
(JIVARA LATTE 40%)
화이트 컴파운드초콜릿 204g
포도씨유 49g

기타

식용금박

COFFEE BEAN CAKE

57g Butter
19g Espresso
115g Whole eggs
46g Egg yolks
16g Honey
77g Hazelnut powder
119g Powdered sugar
75g Cake flour
0.3g Baking powder
4g Ground coffee beans

ALCOHOL SPRAY

200g Coffee liqueur
(KAHLUA)
100g Gold rum
(MONARCH)

COFFEE GANACHE

15g Coffee beans
193g Heavy cream
44g Inverted sugar
193g Dark chocolate
(CARAIBE 66%)
44g Butter
12g Coffee liqueur
(KAHLUA)

CHOCOLATE GLAZE

348g Milk chocolate
(JIVARA LATTE 40%)
204g White compound chocolate
49g Grapeseed oil

OTHER

Edible gold leaves

1
2
3
4-1
4-2
5
6

Process

커피 빈 케이크

1. 녹인 버터(55℃)와 에스프레소(60℃)를 섞는다.
2. 볼에 달걀전란, 달걀노른자, 꿀, 헤이즐넛파우더, 슈거파우더를 넣고 충분히 믹싱한다.
3. 체 친 박력분, 베이킹파우더, 원두커피분말을 넣고 섞는다.
4. 1을 조금씩 넣어가며 섞는다.
5. 커피 빈 모양 실리콘 몰드(Pavoni PX4312)에 40g씩 팬닝한 후 160℃ 오븐에서 약 25분간 굽는다.
6. 구워져 나온 커피 빈 케이크에 알콜 스프레이를 분사한 후 실온에서 완전히 식혀 냉동고에 보관한다.

   * 알콜 스프레이는 커피 리큐어와 골드럼을 섞어 만든다.

COFFEE BEAN CAKE

1. Combine melted butter (55°C) and espresso (60°C).
2. In a bowl, add whole eggs, egg yolks, honey, hazelnut powder, and powdered sugar; mix thoroughly.
3. Sift and add cake flour, baking powder, and ground coffee beans.
4. Gradually add 1 to mix
5. Pour 40 grams into coffee bean-shaped silicone molds (Pavoni PX4312) and bake for about 25 minutes at 160°C.
6. Apply alcohol spray on the baked coffee bean cakes, let cool completely at an ambient temperature, and freeze them.

   * Combine coffee liqueur and gold rum to make the alcohol spray.

7
8
9
BRAUN
10
11
12
13
14
BRAUN
15

커피 가나슈

7. 커피원두는 160℃ 오븐에서 5분간 데워준다.
8. 냄비에 생크림, 전화당을 넣고 60℃ 정도로 가열한다.
9. 볼에 **7**과 **8**을 담고 핸드블렌더로 갈아준다.
10. 수비드 백에 넣어 60℃에서 1시간 동안 수비드 방식으로 인퓨징한다.
11. 체로 걸러 커피원두를 제거한다.
12. 녹인 다크초콜릿(35℃)에 조금씩 나눠 넣어가며 핸드블렌더로 혼합한다.
13. 실온 상태의 버터를 넣고 혼합한다.
14. 커피 리큐어를 넣고 계속해서 혼합한다.
15. 트레이에 균일한 높이로 부어준 후 밀착 랩핑한다. 파이핑하기 적합한 상태가 될 때까지 냉장고에 보관한 후 사용한다.

COFFEE GANACHE

7. Warm the coffee beans for 5 minutes at 160°C.
8. Heat heavy cream and inverted sugar in a saucepan to about 60°C.
9. Grind **7** and **8** in a bowl using an immersion blender.
10. Pour in a sous-vide bag and infuse using the sous-vide method for 1 hour at 60°C.
11. Filter through a sieve to remove the coffee beans.
12. Gradually add melted dark chocolate (35°C), a little bit at a time, and combine with an immersion blender.
13. Add room-temperature butter and mix.
14. Add coffee liqueur and continue to mix.
15. Pour onto a tray in even thickness and cover with plastic wrap, making sure it adheres to the surface. Refrigerate until the texture is consistent enough to use in a piping bag.

CHOCOLATE GLAZE
16
BRAUN
17
FINISH
18
19
21-1
20
Chef inox
21-2
22
23

초콜릿 글레이즈

16. 녹인 초콜릿(40℃)과 포도씨유를 혼합한다.
17. 35℃로 맞춰 사용한다.

마무리

18. 케이크 바닥면에 충전물을 주입할 공간을 만들어준다.
19. 커피 가나슈를 가득 채운다.
20. 스패출러를 이용해 주입구를 매끄럽게 정리한다.
    케이크가 차가워지도록 냉동고에서 최소 30분간 보관한다.
21. 초콜릿 글레이즈로 케이크 바닥면을 코팅한 후 바닥면의 초콜릿 글레이즈가 안정적으로 굳을 때까지 냉동고에 잠시 둔다.
22. 초콜릿 글레이즈로 카카오 케이크 전체를 코팅한다.
23. 완전히 굳기 전에 식용금박으로 장식한다.

CHOCOLATE GLAZE

16. Combine melted chocolate (40°C) and grapeseed oil.
17. Use at 35°C.

FINISH

18. Make a space through the bottom of the cake to inject the filling.
19. Fill with coffee ganache.
20. Use a spatula to organize the opening evenly.
    Keep in the freezer for at least 30 minutes to chill the cakes.
21. Coat the bottom of the cake with the chocolate glaze and freeze until the chocolate glaze becomes stable.
22. Coat the whole cake with the chocolate glaze.
23. Decorate with edible gold leaves before it sets completely.

11 CHESTNUT CAKE

밤 케이크

ingredients - 12 cakes

밤 케이크

버터 105g
우유 17g
헤이즐넛파우더 161g
슈거파우더 102g
밤 페이스트 175g
달걀전란 207g
달걀노른자 83g
꿀 27g
박력분 112g
베이킹파우더 1g

밤 충전물

밤 페이스트 200g
밤 퓌레 100g
다크럼 5g
(NEGRITA)

초콜릿 글레이즈 (실온 유지)

밀크초콜릿 345g
(JIVARA LATTE 40%)
밀크 컴파운드초콜릿 207g
포도씨유 48g

알콜 스프레이

다크럼
(NEGRITA)

기타

보늬밤
참깨

CHESTNUT CAKE

105g Butter
17g Milk
161g Hazelnut powder
102g Powdered sugar
175g Chestnut paste
207g Whole eggs
83g Egg yolks
27g Honey
112g Cake flour
1g Baking powder

CHESTNUT FILLING

200g Chestnut paste
100g Chestnut puree
5g Dark rum
(NEGRITA)

CHOCOLATE GLAZE (MAINTAIN ROOM TEMPERATURE)

345g Milk chocolate
(CARAIBE 66%)
207g Milk compound chocolate
48g Grapeseed oil

ALCOHOL SPRAY

Dark rum
(NEGRITA)

OTHER

Chestnuts in syrup
Sesame seeds

1
2
3
4
5
6
7
8

Process

밤 케이크

1. 헤이즐넛파우더는 160℃ 오븐에서 30분간 구워 식혀 준비한다.
2. 녹인 버터(55℃)와 우유(60℃)를 섞어준다.
3. 푸드프로세서에 헤이즐넛파우더, 슈거파우더, 밤 페이스트, 달걀전란, 달걀노른자, 꿀을 넣고 균일한 상태로 혼합한다.
4. 믹싱볼에 옮긴 후 공기를 충분히 주입하며 믹싱한다.
5. 체 친 박력분, 베이킹파우더를 넣고 섞어준다.
6. **2**를 조금씩 넣어가며 섞어준다.
7. 밤 모양 실리콘 몰드(Silikomart MARRON GLACE 110)에 75g씩 팬닝한 후 160℃ 오븐에서 35분간 굽는다.
8. 구워져 나온 밤 케이크에 알콜 스프레이를 분사한 후 완전히 식혀 냉동고에 보관한다.

CHESTNUT CAKE

1. Roast hazelnut powder for 30 minutes at 160°C and let cool.
2. Combine melted butter (55°C) and milk (60°C).
3. In a food processor, add hazelnut powder, powdered sugar, chestnut paste, whole eggs, egg yolks, and honey; mix thoroughly.
4. Transfer to a mixing bowl, and whip while sufficiently mixing with air.
5. Sift and add cake flour and baking powder.
6. Gradually add **2** little by little.
7. Pour 75 grams into chestnut-shaped silicone molds (Silikomart MARRON GLACE 110) and bake for 35 minutes at 160°C.
8. Apply alcohol spray on the baked chestnut cakes, cool completely, and freeze.

CHESTNUT FILLING
9
10
11
CHOCOLATE GLAZE
(MAINTAIN ROOM TEMPERATURE)
12
13
BRAUN

밤 충전물

9. 푸드프로세서에 밤 페이스트, 밤 퓌레를 넣고 부드럽게 갈아준다.
10. 다크럼을 넣고 가볍게 믹싱한다.
11. 사용하기 전까지 냉장고에 보관한다.

초콜릿 글레이즈 (실온 유지)

12. 녹인 초콜릿(40℃)과 포도씨유를 혼합한다.
13. 35℃로 맞춰 사용한다.

CHESTNUT FILLING

9. Grind chestnut paste and chestnut puree in a food processor into a soft paste.
10. Add dark rum and blend briefly.
11. Refrigerate until use.

CHOCOLATE GLAZE (MAINTAIN ROOM TEMPERATURE)

12. Combine melted chocolate (40°C) and grapeseed oil.
13. Use at 35°C.

FINISH
14
15
16
17
Chef Inox
18
19
20
21-1
21-2

마무리

14. 케이크 바닥면에 충전물을 주입할 공간을 만들어준다.
15. 밤 충전물을 소량 파이핑한다.
16. 보늬밤을 넣고 밤 충전물과 밀착시킨다.
17. 밤 충전물을 가득 채운 후 스패출러로 주입구를 매끄럽게 정리한다. 냉동고에 최소 30분간 보관한다.
18. 초콜릿 글레이즈로 케이크 바닥면을 코팅한다.
19. 바닥면의 초콜릿 글레이즈가 안정적으로 굳을 때까지 냉동고에 잠시 둔다.
20. 초콜릿 글레이즈로 밤 케이크 전체를 코팅한다.
21. 초콜릿 글레이즈가 완전히 굳기 전에 참깨를 묻혀 장식한다.

FINISH

14. Make a space through the bottom of the cake to inject the filling.
15. Pipe a small quantity of chestnut filling.
16. Insert a chestnut in syrup and attach it to the chestnut filling.
17. Fill with chestnut filling and organize the opening with a spatula. Freeze for at least 30 minutes.
18. Coat the bottom of the cake with chocolate glaze.
19. Freeze for a while until the chocolate glaze on the bottom is stable.
20. Coat the whole cake with the chocolate glaze.
21. Dip into sesame seeds before the chocolate glaze sets completely.

12

CARAMEL PECAN FINANCIER

캐러멜 피칸 피낭시에

ingredients - 24 pieces

GLUTEN FREE

캐러멜 피낭시에

설탕A 70g
생크림 70g
헤이즐넛 버터 168g
달걀흰자 208g
꿀 22g
아몬드파우더 131g
고운 쌀가루 70g
옥수수전분 18g
게랑드소금 3g
설탕B 84g
트레할로스 65g

피칸 사블레

물 38g
설탕 128g
피칸 분태 200g

기타

게랑드소금

GLUTEN FREE

CARAMEL FINANCIER

70g Sugar A
70g Heavy cream
168g Hazelnut butter
208g Egg whites
22g Honey
131g Almond powder
70g Fine rice flour
18g Corn starch
3g Guérande salt
84g Sugar B
65g Trehalose

PECAN SABLE

38g Water
128g Sugar
200g Chopped pecans

OTHER

Guérande salt

CARAMEL FINANCIER
1
2
3
4-1
4-2
5
6
7
8
9

Process

캐러멜 피낭시에

1. 냄비에 설탕A를 넣고 캐러멜화시킨다.
2. 골든 캐러멜 색이 되면 뜨겁게 데운 생크림을 조금씩 넣어가며 디글레이즈한다.
3. 완성된 캐러멜 소스는 실온 상태(25~27℃)로 식혀 사용한다.
4. 냄비에 버터 210g을 넣고 가열해 헤이즐넛 버터를 만든다.
5. 달걀흰자, 꿀을 핸드블렌더로 믹싱해 균일한 상태로 만든다.
6. 볼에 체 친 아몬드파우더, 고운 쌀가루, 옥수수전분, 게랑드소금, 설탕B, 트레할로스를 넣고 섞어준다.
7. **5**를 넣고 섞어준다.
8. **3**을 넣고 섞어준다.
9. **4**에서 만든 헤이즐넛 버터(55℃) 168g을 넣고 섞어준다.

CARAMEL FINANCIER

1. Caramelize sugar A in a pot.
2. When it becomes golden brown, gradually add hot cream to deglaze.
3. Cool the finished caramel sauce to ambient temperature (25~27°C) to use.
4. Heat 210 grams of butter in a saucepan to make hazelnut butter.
5. Combine egg whites and honey with an immersion blender to homogenize.
6. In a bowl, sift and add almond powder, fine rice flour, corn starch, Guérande salt, sugar B, and trehalose.
7. Add **5** and mix.
8. Add **3** and mix.
9. Add 168 grams of the hazelnut butter (55°C) from **4** and mix.

PECAN SABLE
10
11
12
13
FINISH
14
15

피칸 사블레

10. 냄비에 물과 설탕을 넣고 118℃로 가열한다.
11. 피칸 분태를 넣고 섞어준다.
12. 설탕이 하얗게 재결정화될 때까지 계속해서 섞어준다.
13. 실팻 위에 넓게 펼친 후 완전히 식혀 사용한다.

마무리

14. 버터를 얇게 칠한 틀에 완성된 반죽을 35g씩 팬닝한다.
15. 완전히 식힌 피칸 사블레를 약 8g씩 토핑한다.
    190℃ 오븐에서 약 15분간 굽고 오븐에서 나오자마자 게랑드소금을 소량씩 토핑한다.
    * 피칸 사블레는 반죽 가장자리까지 골고루 코팅해야 구웠을 때 보기 좋다.

PECAN SABLE

10. Heat water and sugar in a pot to 118°C.
11. Stir in the chopped pecans.
12. Continue to stir until the sugar recrystallizes and turns white.
13. Spread wide over a Silpat and cool completely to use.

FINISH

14. Pour 35 grams of the batter into the lightly greased molds.
15. Top with 8 grams of completely cooled pecan sable.
    Bake for about 15 minutes at 190°C and sprinkle with Guérande salt immediately after removing from the oven.
    * It looks nicer when the pecan sable is sprinkled evenly to the edge.

13 CHOCOLATE & RASPBERRY FINANCIER

초콜릿 & 라즈베리 피낭시에

ingredients - 24 pieces

초콜릿 피낭시에

달걀흰자 250g
아몬드파우더 250g
옥수수전분 42g
카카오파우더 26g
설탕 250g
꿀 50g
헤이즐넛 버터 150g

라즈베리 충전물

냉동 라즈베리 329g
설탕 81g
트레할로스 66g
프람보아즈 리큐어 23g
(DIJON FRAMBOISES)

알콜 스프레이

체리 증류주
(DIJON KIRSCH)

CHOCOLATE FINANCIER

250g Egg whites
250g Almond powder
42g Corn starch
26g Cacao powder
250g Sugar
50g Honey
150g Hazelnut butter

RASPBERRY FILLING

329g Frozen raspberries
81g Sugar
66g Trehalose
23g Framboise liqueur
(DIJON FRAMBOISES)

ALCOHOL SPRAY

Cherry liqueur
(DIJON KIRSCH)

1
2
BRAUN
3
4
5
6-1
6-2

Process

초콜릿 피낭시에

1. 냄비에 버터 188g을 넣고 가열해 붉은 갈색의 헤이즐넛 버터를 만든다.
2. 달걀흰자를 핸드블렌더로 믹싱해 균일한 상태로 만든다.
3. 볼에 체 친 아몬드파우더, 옥수수전분, 카카오파우더, 설탕을 넣고 섞어준다.
4. **3**에 **2**를 넣고 섞어준다.
5. 꿀을 넣고 섞어준다.
6. **1**에서 만든 헤이즐넛 버터(55℃) 150g을 넣고 섞어준다.

CHOCOLATE FINANCIER

1. Heat 188g of butter in a pot to make reddish-brown hazelnut butter.
2. Blend egg white with an immersion blender to homogenize.
3. Mix sifted almond powder, corn starch, cacao powder, and sugar in a bowl.
4. Add **2** into **3** and combine.
5. Add honey and mix.
6. Add the hazelnut butter (55°C) from procedure **1** and mix.

RASPBERRY FILLING
7
8
9
10
11
FINISH
12
13

라즈베리 충전물

7. 냉동 라즈베리, 설탕, 트레할로스를 섞어준다.
8. 냉동 라즈베리가 충분히 해동될 때까지 실온에 12시간 정도 둔다.
9. 냄비에 옮겨 70Brix가 될 때까지 가열한다.
10. 불에서 내려준 후 프람보아즈 리큐어를 넣고 섞어준다.
11. 트레이에 부어 차갑게 식힌 후 사용한다.

마무리

12. 버터를 얇게 칠한 틀에 완성된 반죽을 38g씩 팬닝한다.
13. 반죽 가운데에 라즈베리 충전물을 파이핑한다.
14. 170℃ 오븐에서 15분간 굽는다.
15. 오븐에서 나오자마자 알콜 스프레이를 분사한다.
16. 피낭시에가 완전히 식으면 바닥 부분이 위로 올라오도록 둔다.

RASPBERRY FILLING

7. Combine frozen raspberry, sugar, and trehalose.
8. Let stand at an ambient temperature until the frozen raspberries defrost sufficiently, about 12 hours.
9. Transfer to a saucepan and heat until it becomes 70 Brix.
10. Remove from heat and mix with framboise liqueur.
11. Pour onto a tray and cool completely to use.

FINISH

12. Pour 38 grams of the batter into the lightly greased molds.
13. Pipe raspberry filling in the center.
14. Bake for 15 minutes at 170°C.
15. Spray with alcohol spray immediately after removing from the oven.
16. When the financiers are completely cooled, place them bottom-side up.

14 PISTACHIO & ORANGE FINANCIER

피스타치오 & 오렌지 피낭시에

ingredients - 24 pieces

GLUTEN FREE

피스타치오 & 오렌지 피낭시에

설탕 251g
오렌지제스트 2.5g
달걀흰자 221g
전화당 49g
고운 쌀가루 84g
아몬드파우더 123g
피스타치오파우더 25g
피스타치오 페이스트 37g
헤이즐넛 버터 212g

오렌지 토핑

오렌지 껍질 83g
오렌지 과육 331g
설탕 83g
NH펙틴 4.1g

기타

피스타치오 커넬

GLUTEN FREE

PISTACHIO & ORANGE FINANCIER

251g Sugar
2.5g Orange zest
221g Egg whites
49g Inverted sugar
84g Fine rice flour
123g Almond powder
25g Pistachio powder
37g Pistachio paste
212g Hazelnut butter

ORANGE TOPPING

83g Orange peel
331g Orange pulp
83g Sugar
4.1g Pectin NH

OTHER

Pistachio kernels

1
2
3
4
5
6
7-1
7-2

Process

피스타치오 & 오렌지 피낭시에

1. 냄비에 버터 265g을 넣고 가열해 붉은 갈색의 헤이즐넛 버터를 만든다.
2. 설탕과 오렌지제스트를 미리 혼합해둔다.
3. 달걀흰자와 전화당을 핸드블렌더로 믹싱해 균일한 상태로 만든다.
4. 볼에 체 친 고운 쌀가루, 아몬드파우더, 피스타치오파우더를 넣고 섞어준다.
5. **4**에 **3**을 넣고 섞어준다.
6. 피스타치오 페이스트를 넣고 섞어준다.
7. **1**에서 만든 헤이즐넛 버터 212g을 넣고 섞어준다.

PISTACHIO & ORANGE FINANCIER

1. Heat 265 grams of butter in a pot to make reddish-brown hazelnut butter.
2. In a bowl, combine sugar and orange zest; set aside.
3. Combine egg whites and inverted sugar with an immersion blender to make a homogeneous mixture.
4. In a bowl, combine sifted fine rice flour, almond powder, and pistachio powder.
5. Add **3** into **4** and mix.
6. Add pistachio paste and combine.
7. Add 212 grams of hazelnut butter from procedure 1.

ORANGE TOPPING
8
9
10
11
12
13
14
15
FINISH
16
17

오렌지 토핑

8. 오렌지는 껍질과 과육으로 분리한다. 오렌지 껍질은 끓는 물에 3분씩 3회 데친다.
   * 이때 물은 새 물로 교체한다.
9. 설탕과 NH펙틴을 섞어준다.
10. **8**과 **9**를 섞어준다.
11. 실온에 12시간 정도 두어 과즙이 빠져나오도록 한다.
12. 냄비에 넣고 50~52Brix가 될 때까지 가열한다.
13. 트레이에 부어 차갑게 식힌다.
14. 푸드프로세서에 넣고 균일하게 갈아준다.
15. 사용 전까지 냉장고에 보관한다.

마무리

16. 버터를 얇게 칠한 틀에 완성된 반죽을 40g씩 팬닝한다.
17. 반죽 가운데에 오렌지 토핑을 모양내어 파이핑한 후 피스타치오 커넬을 보기 좋게 토핑한다. 190℃ 오븐에서 12~14분간 굽는다.

ORANGE TOPPING

8. Separate the orange peel and the pulp. Blanch orange peel three times for 3 minutes.
   * Replace the water when blanching.
9. Mix sugar and pectin NH.
10. Combine **8** and **9**.
11. Let stand at an ambient temperature for about 12 hours to extract the juice.
12. Heat in a saucepan until it becomes 50~52 Brix.
13. Pour onto a tray and let cool completely.
14. Blend in a food processor to make a homogenous mixture.
15. Refrigerate until use.

FINISH

16. Pour 40 grams of the batter into lightly greased molds.
17. Pipe orange topping in the center of the batter and arrange pistachio kernels. Bake for 12~14 minutes at 190°C.

15 BLUE CHEESE FINANCIER

블루치즈 피낭시에

ingredients - 24 pieces

블루치즈 피낭시에

달걀흰자 221g
블루치즈 85g
크림치즈 85g
설탕 227g
아몬드파우더 102g
고운 쌀가루 64g
옥수수전분 17g
베이킹파우더 5.2g
게랑드소금 0.8g
헤이즐넛 버터 158g

기타

피칸 분태

BLUE CHEESE FINANCIER

221g Egg whites
85g Blue cheese
85g Cream cheese
227g Sugar
102g Almond powder
64g Fine rice flour
17g Corn starch
5.2g Baking powder
0.8g Guérande salt
158g Hazelnut butter

OTHER

Chopped pecans

1

2

BRAUN

3

4

5

Process

블루치즈 피낭시에

1. 냄비에 버터 197g을 넣고 가열해 붉은 갈색의 헤이즐넛 버터를 만든다.
2. 달걀흰자를 핸드블렌더로 믹싱해 균일한 상태로 만든다.
3. 블루치즈와 크림치즈는 부드럽게 풀어준다.
4. 다른 볼에 설탕, 체 친 아몬드파우더, 고운 쌀가루, 옥수수전분, 베이킹파우더, 게랑드소금을 넣고 섞어준다.
5. **2**를 넣고 섞어준다.

BLUE CHEESE FINANCIER

1. Heat 197 grams of butter in a saucepan to make reddish-brown hazelnut butter.
2. Blend egg whites with an immersion blender to make a homogenous mixture.
3. Soften blue cheese and cream cheese.
4. In a separate bowl, combine sugar, sifted almond powder, fine rice flour, corn starch, baking powder, and Guérande salt.
5. Mix with **2**.

6
7-1
7-2
8
9-1
BRAUN
9-2
FINISH
10
11

6. 1에서 만든 헤이즐넛 버터 158g을 넣고 섞어준다.
7. 부드럽게 푼 블루치즈와 크림치즈에 완성한 피낭시에 반죽 **6**을 일부 혼합한다.
8. 다시 전체 반죽에 넣고 섞어준다.
9. 치즈의 덩어리가 남지 않도록 핸드블렌더로 균일하게 혼합한다.

마무리

10. 버터를 얇게 칠한 틀에 완성된 반죽을 38g씩 팬닝한다.
11. 피칸 분태를 토핑한다.
12. 190℃ 오븐에서 약 12분간 굽는다.
13. 피낭시에가 완전히 식으면 바닥 부분이 위로 올라오도록 둔다.
    * 꿀을 토핑해 즐겨도 좋다.

6. Add 158 grams of hazelnut butter from procedure **1**.
7. Add a portion of the batter from procedure **6** to the softened blue cheese and cream cheese.
8. Add back to the whole batter and mix.
9. Combine with an immersion blender, so no cheese lumps remain.

FINISH

10. Pour 38 grams of the batter into the lightly greased molds.
11. Top with chopped pecans.
12. Bake for about 12 minutes at 190°C.
13. When the financiers are completely cooled, place them bottom-side up.
    * You can drizzle them with honey as well.

16 SMOKED VANILLA MADELEINE

스모크 바닐라 마들렌

ingredients - 24 pieces

NUTS FREE

바닐라 마들렌

버터 167g

정제버터 (콜만 액상버터) 19g

달걀 157g

꿀 35g

설탕 88g

트레할로스 56g

바닐라빈 페이스트 2.9g (NOROHY VANILLA BEAN PASTE - MADAGASCAR)

박력분 205g

베이킹파우더 9.2g

바닐라 시럽

30보메 시럽 200g

물 30g

바닐라 리큐어 115g (DIJON VANILLA)

바닐라빈 깍지 1개

기타

건조시킨 바닐라빈 깍지 (마다가스카르산)

NUTS FREE

VANILLA MADELEINE

167g Butter

19g Clarified butter (Corman Liquid Clarified Butter)

157g Whole eggs

35g Honey

88g Sugar

56g Trehalose

2.9g Vanilla bean paste (NOROHY VANILLA BEAN PASTE - MADAGASCAR)

205g Cake flour

9.2g Baking powder

VANILLA SYRUP

200g 30°B syrup

30g Water

115g Vanilla liqueur (DIJON VANILLA)

1 Vanilla bean pod

OTHER

Dried vanilla bean pod (Madagascar)

1
2
3
4
5

Process

바닐라 마들렌

1. 녹인 버터(55℃)와 정제버터를 혼합한다.
2. 볼에 달걀전란, 꿀, 설탕, 트레할로스, 바닐라빈 페이스트를 넣고 섞어준다.
3. 체 친 박력분, 베이킹파우더를 넣고 섞어준다.
4. 1을 넣고 혼합한다.
5. 냉장고에서 최소 90분간 휴지시킨다.

VANILLA MADELEINE

1. Combine melted butter (55°C) and clarified butter.
2. Combine whole eggs, honey, sugar, trehalose, and vanilla bean paste in a bowl.
3. Sift and add cake flour and baking powder.
4. Combine with 1.
5. Let it rest in a refrigerator for at least 90 minutes.

마무리

6. 틀에 버터를 얇게 칠한다.
7. 틀에 강력분(분량 외)을 체 쳐 도포한 후 뒤집어 여분의 가루를 털어낸다.
8. 틀에 완성된 반죽을 26g씩 팬닝한 후 230℃로 예열된 오븐을 190℃로 낮춰 11분간 굽는다.
9. 구워져 나온 마들렌은 틀에서 분리한 후 바닐라 시럽에 적셔준다.

   * 바닐라 시럽은 모든 재료를 섞어 사용한다.
10. 마들렌이 완전히 식으면 밀폐용기에 담고 건조시킨 마다가스카르산 바닐라빈 깍지를 이용해 1분간 훈연한다.

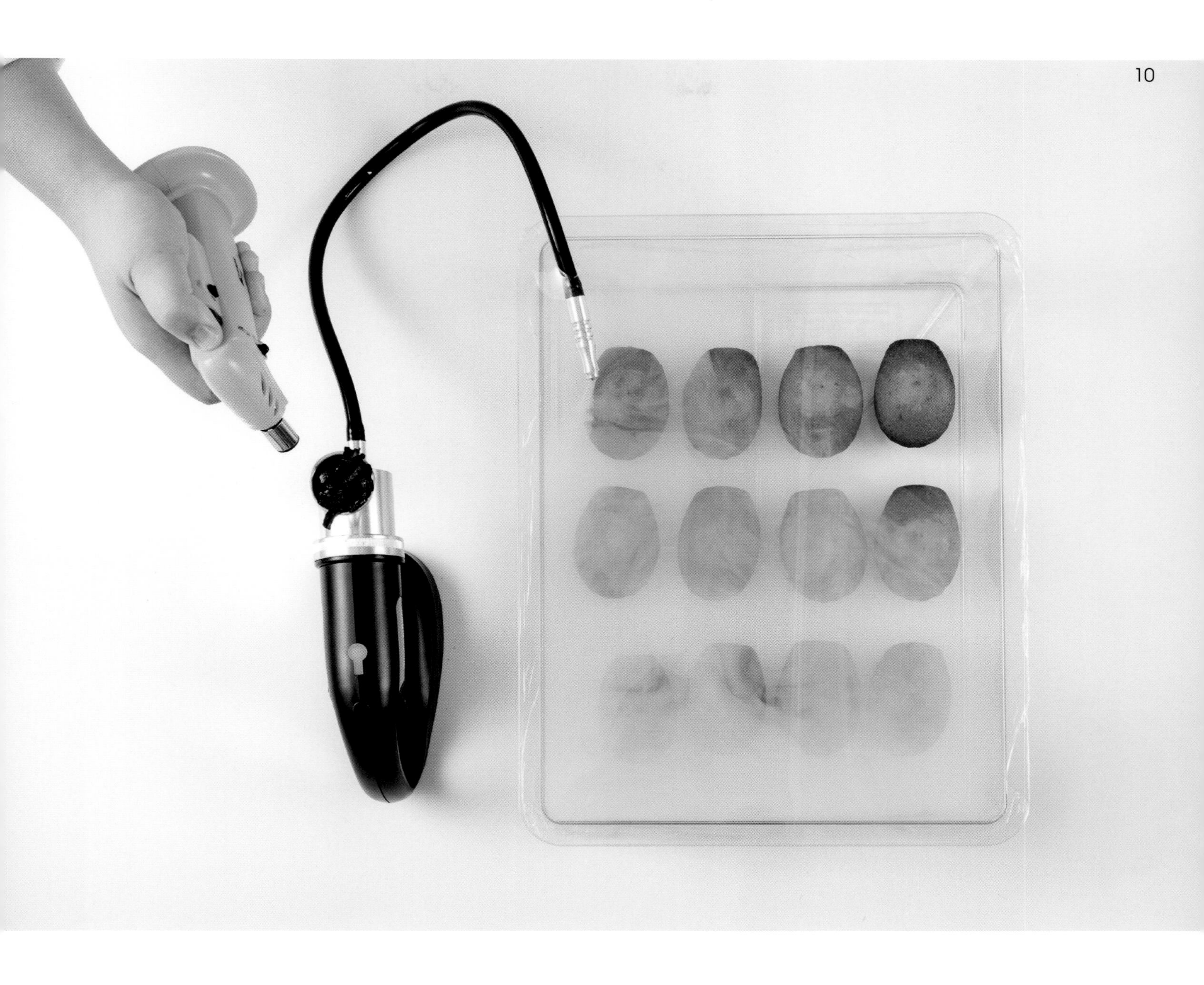
10

FINISH

6. Lightly brush butter in the molds.
7. Dust the mold with bread flour (other than requested) and turn it to remove excess flour.
8. Pipe 26 grams of batter in the molds. Preheat oven to 230°C, then reduce to 190°C and bake for 11 minutes.
9. Remove the baked madeleines from the mold and soak them with vanilla syrup.

   * Combine all the ingredients of vanilla syrup to use.
10. Put the completely cooled madeleines in an air-tight container and smoke for 1 minute using the dried Madagascar vanilla bean.

17 APPLE CRUMBLE MADELEINE

애플 크럼블 마들렌

ingredients - 24 pieces

크럼블

버터 76g
박력분 102g
아몬드파우더 85g
원당 85g

사과 충전물

사과 330g
설탕 66g
바닐라빈 페이스트 0.5g
(NOROHY VANILLA BEAN PASTE - MADAGASCAR)
옥수수전분 9g
사과 증류주 10g
(CALVADOS)

마들렌

버터 150g
정제버터 (콜만 액상버터) 17g
달걀전란 141g
꿀 31g
설탕 79g
트레할로스 50g
바닐라빈 페이스트 2.6g
(NOROHY VANILLA BEAN PASTE - MADAGASCAR)
박력분 183g
베이킹파우더 8.2g

칼바도스 시럽

30보메 시럽 200g
착즙 사과주스 30g
사과 증류주 115g
(CALVADOS)

CRUMBLE

76g Butter
102g Cake flour
85g Almond powder
85g Raw sugar

APPLE FILLING

330g Apples
66g Sugar
0.5g Vanilla bean paste
(NOROHY VANILLA BEAN PASTE - MADAGASCAR)
9g Corn starch
10g Apple liqueur
(CALVADOS)

MADELEINE

150g Butter
17g Clarified butter
(Corman Liquid Clarified Butter)
141g Whole eggs
31g Honey
79g Sugar
50g Trehalose
2.6g Vanilla bean paste
(NOROHY VANILLA BEAN PASTE - MADAGASCAR)
183g Cake flour
8.2g Baking powder

CALVADOS SYRUP

200g 30°B syrup
30g Freshly squeezed apple juice
115g Apple liqueur
(CALVADOS)

CRUMBLE
1
2
3
APPLE FILLING
4
5
6
7
8
9

Process

크럼블

1. 믹싱볼에 실온 상태의 버터, 체 친 박력분, 아몬드파우더, 원당을 넣고 믹싱한다.
2. 한 덩어리로 뭉친 반죽을 체반에 내려 크럼블 형태로 만든 후 냉동고에서 단단하게 굳힌다.
3. 160℃ 오븐에서 약 25분간 구운 후 완전히 식혀 사용한다.

사과 충전물

4. 볼에 사방 0.5cm로 썬 사과, 설탕, 바닐라빈 페이스트를 넣고 섞는다.
5. 실온에 3~4시간 정도 두어 과즙이 나오도록 한다.
6. 냄비에 옮긴 후 옥수수전분을 넣고 혼합한다.
7. 수분이 완전히 없어질 때까지 졸인다.
8. 사과 증류주를 넣고 플람베한다.
9. 트레이에 부어 차갑게 식혀 사용한다.

CRUMBLE

1. In a mixing bowl, beat room temperature butter, sifted cake flour, almond powder, and raw sugar.
2. When the dough comes together, push through a grid to make it into crumbles and freeze until hardened.
3. Bake for about 25 minutes at 160℃ and cool completely to use.

APPLE FILLING

4. Combine apples cut into 0.5 cm cubes, sugar, and vanilla bean paste in a bowl.
5. Let stand at an ambient temperature for 3~4 hours to extract the juice.
6. Transfer into a pot and mix with corn starch.
7. Simmer until all the juice evaporates.
8. Flambé with apple liqueur.
9. Pour onto a tray and cool completely to use.

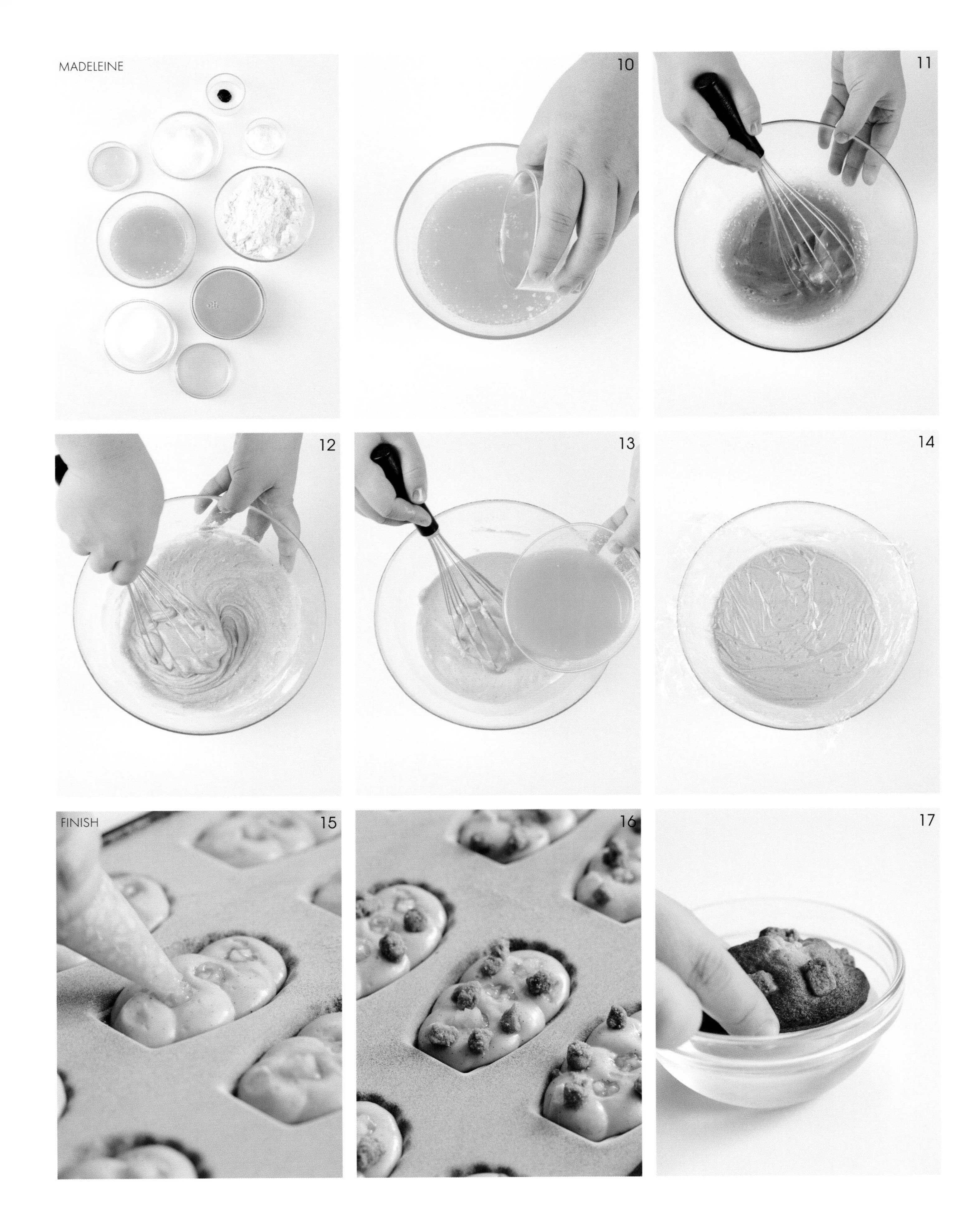
MADELEINE
10
11
12
13
14
FINISH
15
16
17

마들렌

10. 녹인 버터(55℃)와 정제버터를 혼합한다.
11. 다른 볼에 달걀전란, 꿀, 설탕, 트레할로스, 바닐라빈 페이스트를 넣고 섞어준다.
12. 체 친 박력분, 베이킹파우더를 넣고 섞어준다.
13. **10**을 넣고 혼합한다.
14. 냉장고에서 최소 90분간 휴지시킨다.

마무리

15. 틀에 완성된 반죽을 25g씩 팬닝한 후 사과 충전물을 5g씩 파이핑한다 .
    * 틀에 밀가루(강력분)를 체 쳐 도포한 후 뒤집어 여분의 가루를 털어낸다.
16. 크럼블을 5g씩 올린 후 230℃로 예열된 오븐을 190℃로 낮춰 11분간 굽는다.
17. 구워져 나온 마들렌은 틀에서 분리한 후 칼바도스 시럽에 적셔준다.
    * 칼바도스 시럽은 모든 재료를 섞어 사용한다.

MADELEINE

10. Combine melted butter (55°C) and clarified butter.
11. Combine whole eggs, honey, sugar, trehalose, and vanilla bean paste in a separate bowl.
12. Sift and add cake flour and baking powder.
13. Combine with **10**.
14. Let it rest in a refrigerator for at least 90 minutes.

FINISH

15. Pipe 25 grams of batter in the molds and pipe 5 grams of apple filling.
    * Dust the mold with flour (bread flour) and turn it to remove excess flour.
16. Top with 5 grams of crumbles. Preheat oven to 230°C, then reduce to 190°C and bake for 11 minutes.
17. Remove the baked madeleines from the mold and soak them with Calvados syrup.
    * Combine all the ingredients of Calvados syrup to use.

18 KAFFIR LIME & BASIL MADELEINE

카피르 라임 & 바질 마들렌

ingredients - 24 pieces

카피르 라임 & 바질 마들렌

버터 165g
정제버터 19g
(콜만 액상버터)
설탕 87g
트레할로스 55g
카피르 라임제스트 3.8g
바질 5.7g
달걀전란 156g
꿀 34g
박력분 203g
베이킹파우더 9.1g

패션푸르트 시럽

30보메 시럽 200g
라임주스 30g
패션프루트 리큐어 115g
(DIJON FRUIT DE LA PASSION)

폰당

슈거파우더 250g
라임주스 60g
포도씨유 10g
바질 2g

기타

카피르 라임제스트

KAFFIR LIME & BASIL MADELEINE

165g Butter
19g Clarified butter
(Corman Liquid Clarified Butter)
87g Sugar
55g Trehalose
3.8g Kaffir lime zest
5.7g Basil
156g Whole eggs
34g Honey
203g Cake flour
9.1g Baking powder

PASSION FRUIT SYRUP

200g 30°B syrup
30g Lime juice
115g Passion fruit liqueur
(DIJON FRUIT DE LA PASSION)

FONDANT

250g Powdered sugar
60g Lime juice
10g Grapeseed oil
2g Basil

OTHER

Kaffir lime zest

KAFFIR LIME & BASIL MADELEINE
1
2
3
4
5
6
FONDANT
7

Process

카피르 라임 & 바질 마들렌

1. 녹인 버터(55℃)와 정제버터를 혼합한다.
2. 푸드프로세서에 설탕, 트레할로스, 카피르 라임제스트, 바질을 넣고 갈아준다.
3. 볼에 **2**와 달걀전란, 꿀을 넣고 섞어준다.
4. 체 친 박력분, 베이킹파우더를 넣고 섞어준다.
5. **1**을 넣고 혼합한다.
6. 냉장고에서 최소 90분간 휴지시킨다.

폰당

7. 푸드프로세서에 모든 재료를 넣고 믹싱한다.

KAFFIR LIME & BASIL MADELEINE

1. Mix melted butter (55℃) and clarified butter.
2. Blend sugar, trehalose, kaffir lime zest, and basil with a food processor.
3. Mix **2** with whole eggs and honey in a bowl.
4. Add sifted cake flour and baking powder.
5. Add **1** to combine.
6. Refrigerate for at least 90 minutes.

FONDANT

7. Mix all the ingredients in a food processor.

마무리

8. 반죽을 26g씩 팬닝한 후 230℃로 예열된 오븐을 190℃로 낮춰 11분간 굽는다.
   * 틀에 밀가루(강력분)를 체 쳐 도포한 후 뒤집어 여분의 가루를 털어낸다.
9. 구워져 나온 마들렌은 틀에서 분리한 후 패션푸르트 시럽에 적셔준다.
   * 패션푸르트 시럽은 모든 재료를 섞어 사용한다.
10. 마들렌이 완전히 식으면 폰당을 발라준 후 170℃ 오븐에서 1분간 건조시킨다.
11. 라임제스트를 뿌린다.

FINISH

8. Pipe 26 grams of batter in the molds. Preheat oven to 230°C, then reduce to 190°C and bake for 11 minutes.

   * Dust the mold with flour (bread flour) and turn it to remove excess flour.

9. Remove the baked madeleines from the mold and soak them with passion fruit syrup.

   * Combine all the ingredients of passion fruit syrup to use.

10. After completely cooling the madeleines, brush them with fondant and dry them for 1 minute at 170°C.

11. Sprinkle with lime zest.

19 RASPBERRY MADELEINE

라즈베리 마들렌

ingredients - 24 pieces

라즈베리 충전물

냉동 라즈베리 172g
라즈베리 퓌레 140g
설탕 62g
옥수수전분 17g
체리 증류주 9g
(DIJON KIRSCH)

마들렌

버터 168g
정제버터 19g
(콜만 액상버터)
달걀전란 156g
꿀 35g
설탕 89g
트레할로스 56g
박력분 206g
베이킹파우더 9.2g

체리 시럽

30보메 시럽 200g
레몬주스 30g
체리 증류주 115g
(DIJON KIRSCH)

RASPBERRY FILLING

172g Frozen raspberries
140g Raspberry puree
62g Sugar
17g Corn Starch
9g Cherry liqueur
(DIJON KIRSCH)

MADELEINE

168g Butter
19g Clarified butter
(Corman Liquid Clarified Butter)
156g Whole eggs
35g Honey
89g Sugar
56g Trehalose
206g Cake flour
9.2g Baking powder

CHERRY SYRUP

200g 30°B syrup
30g Lemon juice
115g Cherry liqueur
(DIJON KIRSCH)

RASPBERRY FILLING
1
2
3
4
5
6

Process

라즈베리 충전물

1. 볼에 냉동 라즈베리, 라즈베리 퓌레, 설탕을 넣고 섞어준다.
2. 실온에 12시간 정도 두고 냉동 라즈베리를 충분히 해동시킨다.
3. 냄비에 옮긴 후 옥수수전분을 넣고 혼합한다.
4. 수분이 충분히 날아갈 때까지 졸인다.
5. 체리 증류주를 넣고 플람베한다.
6. 트레이에 부어 차갑게 식힌 후 사용한다.

RASPBERRY FILLING

1. Mix frozen raspberry, raspberry puree, and sugar in a bowl.
2. Let stand at an ambient temperature for about 12 hours to defrost frozen raspberries sufficiently.
3. Transfer to a pot and combine with corn starch.
4. Simmer sufficiently until all the juice evaporates.
5. Flambé with cherry liqueur.
6. Pour onto a tray and cool completely to use.

7
8
9
10
11

마들렌

7. 녹인 버터(55℃)와 정제버터를 혼합한다.
8. 다른 볼에 달걀전란, 꿀, 설탕, 트레할로스를 넣고 섞어준다.
9. 체 친 박력분, 베이킹파우더를 넣고 섞어준다.
10. **7**을 넣고 섞어준다.
11. 냉장고에서 최소 90분간 휴지시킨다.

MADELEINE

7. Mix melted butter (55°C) and clarified butter.
8. Combine whole eggs, honey, sugar, and trehalose in a separate bowl.
9. Add sifted cake flour and baking powder to mix.
10. Add **7** to mix.
11. Refrigerate for at least 90 minutes.

마무리

12. 반죽을 23g씩 팬닝한 후 라즈베리 충전물을 5g씩 파이핑한다.

    * 틀에 밀가루(강력분)를 체 쳐 도포한 후 뒤집어 여분의 가루를 털어낸다.

13. 뾰족한 도구를 이용해 라즈베리 충전물을 마블 형태로 만든다.

14. 230℃로 예열된 오븐을 190℃로 낮춰 11분간 굽는다.

15. 구워져 나온 마들렌은 틀에서 분리한 후 체리 시럽에 적셔준다.

    * 체리 시럽은 모든 재료를 섞어 만든다.

FINISH

12. Pipe 23 grams of batter in the mold and pipe 5 grams of raspberry filling.

    * Dust the mold with flour (bread flour) and turn it to remove excess flour.

13. Use a pointy tool to give the filling a marble effect.

14. Preheat oven to 230°C, then reduce to 190°C and bake for 11 minutes.

15. Remove the baked madeleines from the mold and soak with cherry syrup.

    * Combine all the ingredients of cherry syrup to use.

20 BURNT VANILLA CANNELE

번트 바닐라 까눌레

ingredients - 10 pieces

NUTS FREE

스모크 바닐라 가나슈

화이트초콜릿 247g (OPALYS 33%)
카카오버터 31g
전화당 15g
생크림 154g
번트 바닐라빈 2.5개
버터 46g
바닐라 리큐어 6g (DIJON VANILLA)

번트 바닐라 까눌레

번트 바닐라빈 2.5개
슈거파우더 187g
우유 389g
버터 62g
T55밀가루 109g
달걀전란 78g
달걀노른자 31g
다크럼 78g (NEGRITA)

기타

건조시킨 바닐라빈 깍지
식용금박
천연 밀랍

NUTS FREE

SMOKED VANILLA GANACHE

247g White chocolate (OPALYS 33%)
31g Cacao butter
15g Inverted sugar
154g Heavy cream
2.5 Burnt vanilla beans
46g Butter
6g Vanilla liqueur (DIJON VANILLA)

BURNT VANILLA CANNELE

2.5 Burnt Vanilla beans
187g Powdered sugar
389g Milk
62g Butter
109g T55 flour
78g Whole eggs
31g Egg yolks
78g Dark rum (NEGRITA)

OTHER

Dried vanilla bean pod
Edible gold leaves
Natural beeswax

SMOKED VANILLA GANACHE
1
2
3
4
5
6
7
8
9
BRAUN
BRAUN

Process

스모크 바닐라 가나슈

1. 볼에 녹인 화이트초콜릿(55℃)과 녹인 카카오버터(55℃)를 넣고 섞어준다.
2. 냄비에 전화당, 생크림, 잘게 자른 번트 바닐라빈을 넣고 60℃로 가열한다.
3. **2**에 번트 바닐라빈을 넣고 핸드블렌더로 갈아준 후 5분 정도 그대로 두어 인퓨징한다.
4. 체에 거른다.
5. **1**에 **4**를 조금씩 넣어가며 섞어준다.
6. 실온 상태의 버터를 넣고 핸드블렌더로 혼합한다.
7. 바닐라 리큐어를 넣고 계속해서 혼합한다.
8. 건조시킨 바닐라빈 깍지를 이용해 5분간 훈연한다.
9. 밀착 랩핑한 후 파이핑하기 적합한 상태가 될 때까지 냉장고에 보관해 사용한다.

SMOKED VANILLA GANACHE

1. Combine melted white chocolate (55°C) and cacao butter (55°C) in a bowl.
2. Heat inverted sugar, heavy cream, and finely chopped burnt vanilla beans to 60°C in a pot.
3. Grind the burnt vanilla bean mixture (**2**) with an immersion blender and let stand to infuse for about 5 minutes.
4. Filter through a sieve.
5. Gradually add **4** into **1** little by little to mix.
6. Add room-temperature butter and combine using an immersion blender.
7. Add vanilla liqueur and continue to mix.
8. Smoke for 5 minutes with the dried vanilla bean pod.
9. Cover with plastic wrap, making sure it adheres to the surface. Refrigerate until the texture is consistent enough to use in a piping bag.

10
11
12
13
14
15
16
17

번트 바닐라 까눌레

10. 유산지를 깐 팬에 바닐라빈을 올리고 130℃ 오븐에서 약 1시간 구워 번트 바닐라빈을 만든다.
11. 푸드프로세서에 잘게 자른 번트 바닐라빈, 슈거파우더를 넣고 곱게 간다.
12. 냄비에 우유, 버터를 넣고 80℃로 가열한다.
13. 볼에 **11**과 T55밀가루를 넣고 섞어준다.
14. 달걀전란, 달걀노른자를 넣고 섞어준다.
15. **12**를 여러 번 나눠 넣어가며 혼합한다.
16. 체에 거른 후 하룻밤 휴지시킨다.
17. 굽기 직전 다크럼을 넣고 섞어준다.

BURNT VANILLA CANNELE

10. Place vanilla beans on a baking tray lined with parchment paper. Bake for about 1 hour at 130°C to make burnt vanilla beans.
11. Grind finely chopped burnt vanilla beans and powdered sugar in a food processor to make a fine powder.
12. Heat milk and butter in a saucepan to 80°C.
13. Mix **11** and T55 flour in a bowl.
14. Add whole eggs and egg yolks.
15. Gradually add **12** little by little to mix.
16. Filter through a sieve and let rest overnight.
17. Mix with dark rum just before baking.

마무리

18. 오븐에서 뜨겁게 데운 까눌레 틀에 녹인 천연 밀랍을 부어준다.
19. 틀을 뒤집어 여분의 밀랍을 털어낸 후 그대로 식혀준다.
20. 밀랍을 코팅한 까눌레 틀에 반죽을 90% 팬닝한다.
21. 230℃ 오븐에서 10분간 굽고, 180℃로 낮춰 30분간 더 굽는다.

FINISH

18. Pour melted natural beeswax into cannele molds warmed hot in the oven.
19. Turn the molds over to remove excess beeswax and leave to cool as is.
20. Fill 90% of the beeswax-coated mold with the batter.
21. Bake for 10 minutes at 230°C, reduce to 180°C, and bake for 30 minutes more.

22. 충분히 식힌 까눌레는 원통 형태의 도구를 이용해 충전물을 채울 공간을 만들어준다.

23. 스모크 바닐라 가나슈를 가득 채운 후 윗부분이 봉긋하게 올라오도록 모양내어 파이핑한다.

24. 식용금박을 올려 장식한다.

22. After the canneles are completely cooled, use a cylinder-shaped tool (such as an apple corer) to make space for the filling.
23. Fill the space with smoked vanilla ganache and pipe on top of canneles to decorate.
24. Top with edible gold to finish.

21 ROASTED CORN CANNELE

구운 옥수수 까눌레

ingredients - 10 pieces

NUTS FREE

구운 옥수수 까눌레

우유 405g
버터 57g
콘푸레이크 (low sugar) 28g
T55밀가루 100g
슈거파우더 171g
소금 4.3g
달걀전란 71g
달걀노른자 28g
골드럼 71g
(MONARCH)

구운 옥수수 가나슈

생크림 135g
전화당 46g
옥수수초콜릿 237g
(리퍼블리카 델 카카오 WHITE CHOCOLATE - ROASTED CORN ECUADOR 33%)
카카오버터 30g
버터 46g
골드럼 7g
(MONARCH)

기타

옥수수수염
식용유
게랑드소금
천연 밀랍

NUTS FREE

ROASTED CORN CANNELE

405g Milk
57g Butter
28g Cornflakes cereal (low sugar)
100g T55 flour
171g Powdered sugar
4.3g Salt
71g Whole eggs
28g Egg yolks
71g Gold rum (MONARCH)

ROASTED CORN GANACHE

135g Heavy cream
46g Inverted sugar
237g Corn chocolate
(REPUBLICA DEL CACAO WHITE CHOCOLATE - ROASTED CORN ECUADOR 33%)
30g Cacao butter
46g Butter
7g Gold rum (MONARCH)

OTHER

Corn husks
Cooking oil
Guérande salt
Natural beeswax

ROASTED CORN CANNELE
1
2
3
4
5
6
7
8

Process

구운 옥수수 까눌레

1. 냄비에 우유, 버터를 넣고 80℃로 가열한다.
2. 콘푸레이크를 넣고 5분간 인퓨징한다.
3. 체에 거른다.
4. 볼에 체 친 T55밀가루, 슈거파우더, 소금을 섞어준다.
5. 달걀전란, 달걀노른자를 넣고 섞어준다.
6. **2**를 여러 번 나눠 넣어가며 혼합한다.
7. 체에 거른다.
8. 하룻밤 휴지시킨다.

ROASTED CORN CANNELE

1. Heat milk and butter to 80°C in a pot.
2. Add cornflakes cereal and let infuse for 5 minutes.
3. Filter through a sieve.
4. Sift T55 flour, powdered sugar, and salt in a bowl.
5. Add whole eggs and egg yolks to mix.
6. Gradually add **2** little by little to combine.
7. Filter through a sieve.
8. Refrigerate overnight.

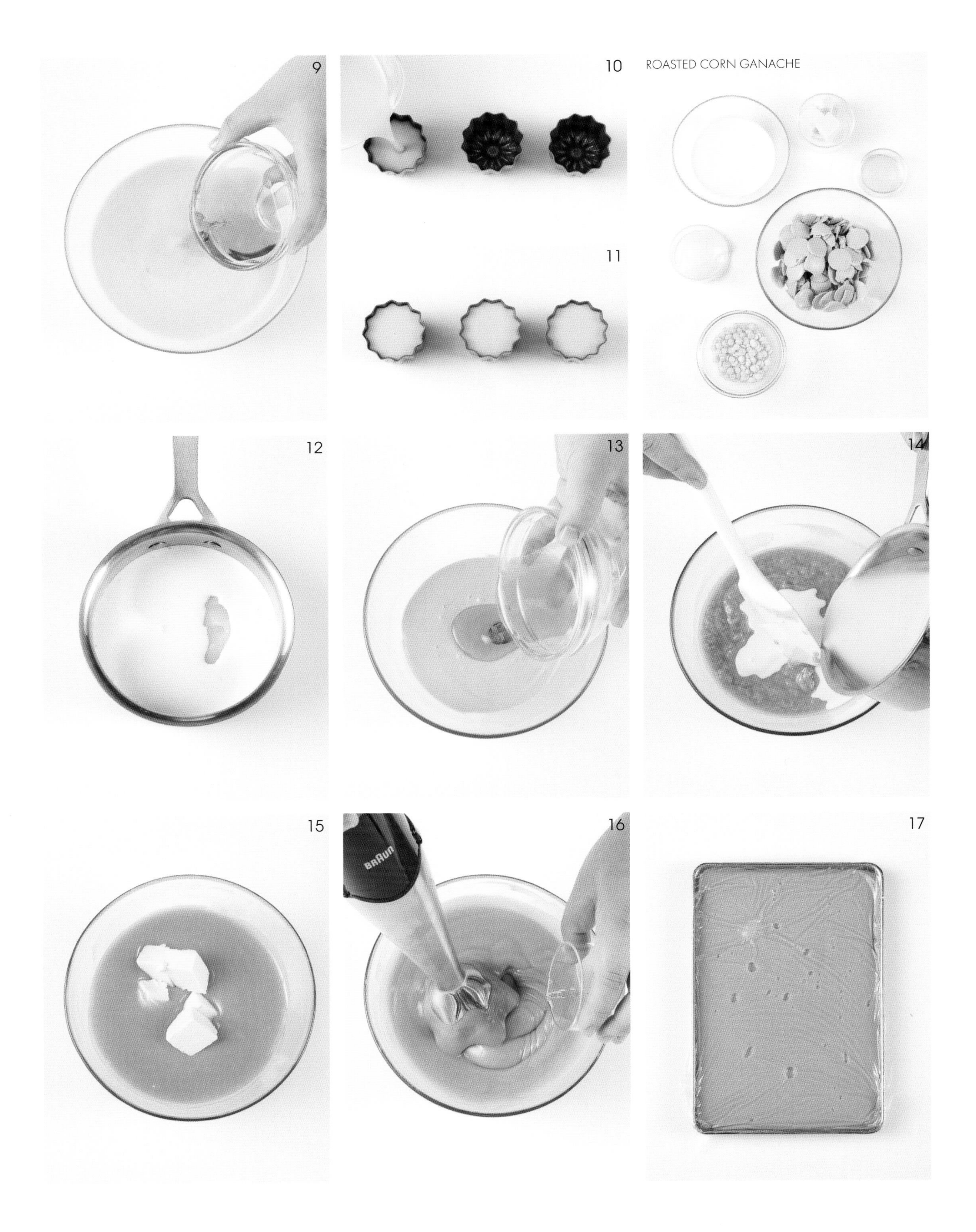
9
10
ROASTED CORN GANACHE
11
12
13
14
15
BRAUN
16
17

9. 굽기 직전 골드럼을 섞어준다.
10. 천연 밀랍을 코팅한 틀에 반죽을 90% 팬닝한다.
11. 230℃ 오븐에서 10분간 굽고, 180℃로 낮춰 30분간 더 굽는다.

구운 옥수수 가나슈

12. 냄비에 생크림, 전화당을 넣고 50~60℃로 가열한다.
13. 볼에 녹인 옥수수초콜릿(35℃)과 녹인 카카오버터(55℃)에 붓고 핸드블렌더로 혼합한다.
14. **13**에 **12**를 나눠 넣어가며 혼합한다.
15. 버터를 넣고 혼합한다.
16. 골드럼을 넣고 혼합한다.
17. 트레이에 부어 차갑게 식힌 후 파이핑하기 적합한 상태가 될 때까지 냉장고에 보관한 후 사용한다.

9. Mix with gold rum just before baking.
10. Fill 90% of the beeswax-coated mold with the batter.
11. Bake for 10 minutes at 230°C, reduce to 180°C, and bake for 30 minutes more.

ROASTED CORN GANACHE

12. Heat heavy cream and inverted sugar in a saucepan to 50~60°C.
13. Combine melted corn chocolate (35°C) and cacao butter (55°C) in a bowl with an immersion blender.
14. Gradually add **12** little by little into **13** to mix.
15. Add butter to combine.
16. Add rum and mix.
17. Pour onto a tray and refrigerate until the texture is consistent enough to use in a piping bag.

마무리

18. 충분히 식힌 까눌레는 원통 형태의 도구를 이용해 충전물을 채울 공간을 만들어준다.

19. 구운 옥수수 가나슈를 가득 채운 후 윗부분이 봉긋하게 올라오도록 모양내어 파이핑한다.

20. 옥수수수염을 올린다.

    * 옥수수수염은 기름에 살짝 튀긴 후 기름기를 제거해 사용한다.

21. 게랑드소금을 토핑한다.

FINISH

18. After the canneles are completely cooled, use a cylinder-shaped tool (such as an apple corer) to make space for the filling.
19. Fill the space with roasted corn ganache and pipe on the canneles to decorate.
20. Top with corn husk.

    * Lightly fry corn husk in oil and drain excess oil to use.

21. Sprinkle Guérande salt to finish.

22 BASIL & RASPBERRY CANNELE

바질 & 라즈베리 까눌레

ingredients - 10 pieces

NUTS FREE

바질 까눌레

우유 385g
버터 62g
슈거파우더 185g
T55밀가루 108g
달걀전란 77g
달걀노른자 31g
바질 11g
체리 증류주 77g
(DIJON KIRSCH)

라즈베리 충전물

NH펙틴 5g
설탕 62g
냉동 라즈베리 146g
라즈베리 퓌레 146g
레몬주스 50g
전화당 32g
트레할로스 45g
프람보아즈 리큐어 15g
(DIJON FRAMBOISES)

기타

천연 밀랍
바질

NUTS FREE

BASIL CANNELE

385g Milk
62g Butter
185g Powdered sugar
108g T55 flour
77g Whole eggs
31g Egg yolks
11g Basil
77g Cherry liqueur
(DIJON KIRSCH)

RASPBERRY FILLING

5g Pectin NH
62g Sugar
146g Frozen raspberries
146g Raspberry puree
50g Lemon juice
32g Inverted sugar
45g Trehalose
15g Framboise liqueur
(DIJON FRAMBOISES)

OTHER

Natural beeswax
Basil

BASIL CANNELE

Process

바질 까눌레

1. 냄비에 우유, 버터를 넣고 80℃로 가열한다.
2. 볼에 체 친 슈거파우더, T55밀가루를 넣고 섞어준다.
3. 달걀전란, 달걀노른자를 넣고 섞어준다.
4. **1**을 여러 번 나눠 넣어가며 혼합한다.
5. 반죽의 250g을 덜어낸 후 얼음물이 담긴 볼에 받쳐 28~30℃로 식힌다.
6. **5**에 바질을 넣고 핸드블렌더로 갈아준다.
7. 5분간 그대로 두어 인퓨징한다.
8. **7**을 남은 반죽에 부어 혼합한다.

BASIL CANNELE

1. Heat milk and butter in a pot to 80°C.
2. Sift and mix powdered sugar and T55 flour in a bowl.
3. Add and mix whole eggs and egg yolks.
4. Gradually add **1** little by little to combine.
5. Remove 250 grams of batter. Cool the removed batter over an ice bath to 28~30°C.
6. Add basil to **5** and grind with an immersion blender.
7. Let stand for 5 minutes to infuse.
8. Pour **7** into the remaining batter and combine.

9
10
11
12
RASPBERRY FILLING
13
14
15
16
17
18

9. 반죽을 체에 거른 후 하룻밤 휴지시킨다.
10. 굽기 직전 체리 증류주를 섞어준다.
11. 천연 밀랍을 코팅한 틀에 반죽을 90% 팬닝한다.
12. 230℃ 오븐에서 약 10분간 굽고, 180℃로 낮춰 30분간 더 굽는다.

라즈베리 충전물

13. NH펙틴과 설탕을 섞어준다.
14. 볼에 **13**과 냉동 라즈베리, 라즈베리 퓌레, 레몬주스, 전화당, 트레할로스를 넣고 섞어준다.
15. 냉동 라즈베리가 충분히 해동될 때까지 실온에 12시간 정도 둔다.
16. **15**를 냄비에 넣고 45Brix가 될 때까지 가열한다.
17. 불에서 내린 후 프람보아즈 리큐어를 넣고 섞어준다.
18. 트레이에 부어 차갑게 식힌 후 부드럽게 풀어 사용한다.

9. Filter the batter through a sieve and let rest overnight.
10. Mix with cherry liqueur just before baking.
11. Fill 90% of the beeswax-coated mold with the batter.
12. Bake for 10 minutes at 230°C, reduce to 180°C, and bake for 30 minutes more.

RASPBERRY FILLING

13. Mix pectin NH with sugar.
14. Combine **13** with frozen raspberries, raspberry puree, lemon juice, inverted sugar, and trehalose in a bowl.
15. Let stand for about 12 hours to defrost frozen raspberries sufficiently.
16. Pour **15** into a saucepan and heat until it reaches 45 Brix.
17. Remove from heat and mix with framboise liqueur.
18. Pour onto a tray, cool completely, and soften to use.

마무리

19. 충분히 식힌 까눌레는 원통 형태의 도구를 이용해 충전물을 채울 공간을 만들어준다.
20. 라즈베리 충전물을 가득 채운 후 윗부분이 봉긋하게 올라오도록 모양내어 파이핑한다.
21. 신선한 바질을 올려 장식한다.

FINISH

19. After the canneles are completely cooled, use a cylinder-shaped tool (such as an apple corer) to make space for the filling.
20. Fill the space with raspberry filling and pipe on the canneles to decorate.
21. Top with fresh basil to finish.

23 ORANGE & GINGER CANNELE

오렌지 & 생강 까눌레

ingredients - 10 pieces

NUTS FREE

오렌지 & 생강 까눌레

우유 393g
버터 63g
오렌지제스트 4.2g
슈거파우더 189g
T55밀가루 94g
생강파우더 3.9g
달걀전란 79g
달걀노른자 31g
오렌지 리큐어 79g
(Grand Marnier)

오렌지 충전물

NH펙틴 8.2g
설탕 58g
전처리한 오렌지 165g
오렌지주스 79g
패션프루트 퓌레 89g
레몬주스 27g
꿀 31g
트레할로스 40g
생강즙 3g

기타

생강 콩피
천연 밀랍

NUTS FREE

ORANGE & GINGER CANNELE

393g Milk
63g Butter
4.2g Orange zest
189g Powdered sugar
94g T55 flour
3.9g Ginger powder
79g Whole eggs
31g Egg yolks
79g Orange liqueur
(Grand Marnier)

ORANGE FILLING

8.2g Pectin NH
58g Sugar
165g Pre-treated orange
79g Orange juice
89g Passion fruit puree
27g Lemon juice
31g Honey
40g Trehalose
3g Ginger juice

OTHER

Ginger confit
Natural beeswax

ORANGE & GINGER CANNELE
1
2
3
4
5
6
7
8
9
10

Process

오렌지 & 생강 까눌레

1. 냄비에 우유, 버터를 넣고 80℃로 가열한다.
2. 불에서 내린 후 오렌지제스트를 넣고 약 5분간 인퓨징한다.
3. 볼에 슈거파우더, T55밀가루, 생강파우더를 넣고 섞어준다.
4. 달걀전란, 달걀노른자를 넣고 섞어준다.
5. **2**를 여러 번 나눠 넣으면서 혼합한다.
6. 체에 거른다.
7. 하룻밤 휴지시킨다.
8. 굽기 직전 오렌지 리큐어를 섞어준다.
9. 천연 밀랍을 코팅한 틀에 반죽을 90% 팬닝한다.
10. 230℃ 오븐에서 약 10분간 굽고, 180℃로 낮춰 30분간 더 굽는다.

ORANGE & GINGER CANNELE

1. Heat milk and butter in a saucepan to 80°C.
2. Remove from heat, add orange zest and let infuse for 5 minutes.
3. Combine powdered sugar, T55 flour, and ginger powder in a bowl.
4. Add whole eggs and egg yolks and mix.
5. Gradually add **2** little by little to mix.
6. Filter through a sieve.
7. Refrigerate overnight.
8. Mix with orange liqueur just before baking.
9. Fill 90% of the beeswax-coated mold with the batter.
10. Bake for 10 minutes at 230°C, reduce to 180°C, and bake for 30 minutes more.

ORANGE FILLING
11
12
13
14
15
16
17
18
19

오렌지 충전물

11. 끓는 물에 오렌지를 넣고 껍질이 완전히 부드러워질 때까지 약 1시간 30분간 삶는다.
    * 이때 15분마다 새 물로 교체한다.
12. 물기를 제거한 후 잘게 자른다.
13. NH펙틴과 설탕을 섞어준다.
14. 냄비에 **12**의 전처리한 오렌지 165g, 오렌지주스, 패션푸르트 퓌레, 레몬주스, 꿀, 트레할로스를 넣고 가열한다.
15. 40℃가 되면 **13**을 넣고 혼합한 후 45Brix가 될 때까지 가열한다.
16. 핸드블렌더로 곱게 갈아준다.
17. 트레이에 부어 차갑게 식힌다.
18. 생강즙을 넣고 섞는다.
19. 부드럽게 풀어 사용한다.

ORANGE FILLING

11. Blanch orange in boiling water until the skin becomes completely soft, about 1 hour and 30 minutes.
    * Replace the water every 15 minutes.
12. Drain water and chop finely.
13. Mix pectin NH with sugar.
14. Heat 165 grams of pre-treated orange from **12**, orange juice, passion fruit puree, lemon juice, honey, and trehalose in a pot.
15. Add **13** when it's 40°C and heat until it reaches 45 Brix.
16. Grind finely with an immersion blender.
17. Pour onto a tray and cool completely.
18. Add ginger juice and mix.
19. Soften to use.

마무리

20. 충분히 식힌 까눌레는 원통 형태의 도구를 이용해 충전물을 채울 공간을 만들어준다.

21. 오렌지 충전물을 가득 채운 후 윗부분이 봉긋하게 올라오도록 모양내어 파이핑한다.

22. 생강 콩피를 토핑한다.

FINISH

20. After the canneles are completely cooled, use a cylinder-shaped tool (such as an apple corer) to make space for the filling.
21. Fill the space with orange filling and pipe on the canneles to decorate.
22. Top with ginger confit to finish.

24 WHITE SOYBEAN PASTE COOKIE

백된장 쿠키

ingredients - 12 cookies

NUTS FREE

쿠키

버터 143g
백된장 50g
바닐라빈 페이스트 0.5g
(Norohy)
원당 114g
무스코바도 설탕 143g
달걀전란 86g

박력분 286g
베이킹파우더 2.9g
베이킹소다 2.9g
다크초콜릿 172g
(GUANAJA 70%)

기타

게랑드소금

NUTS FREE

COOKIE

143g Butter
50g White soybean paste
0.5g Vanilla bean paste
(Norohy)
114g Raw sugar
143g Muscovado sugar
86g Whole eggs

286g Cake flour
2.9g Baking powder
2.9g Baking soda
172g Dark chocolate
(GUANAJA 70%)

OTHER

Guérande salt

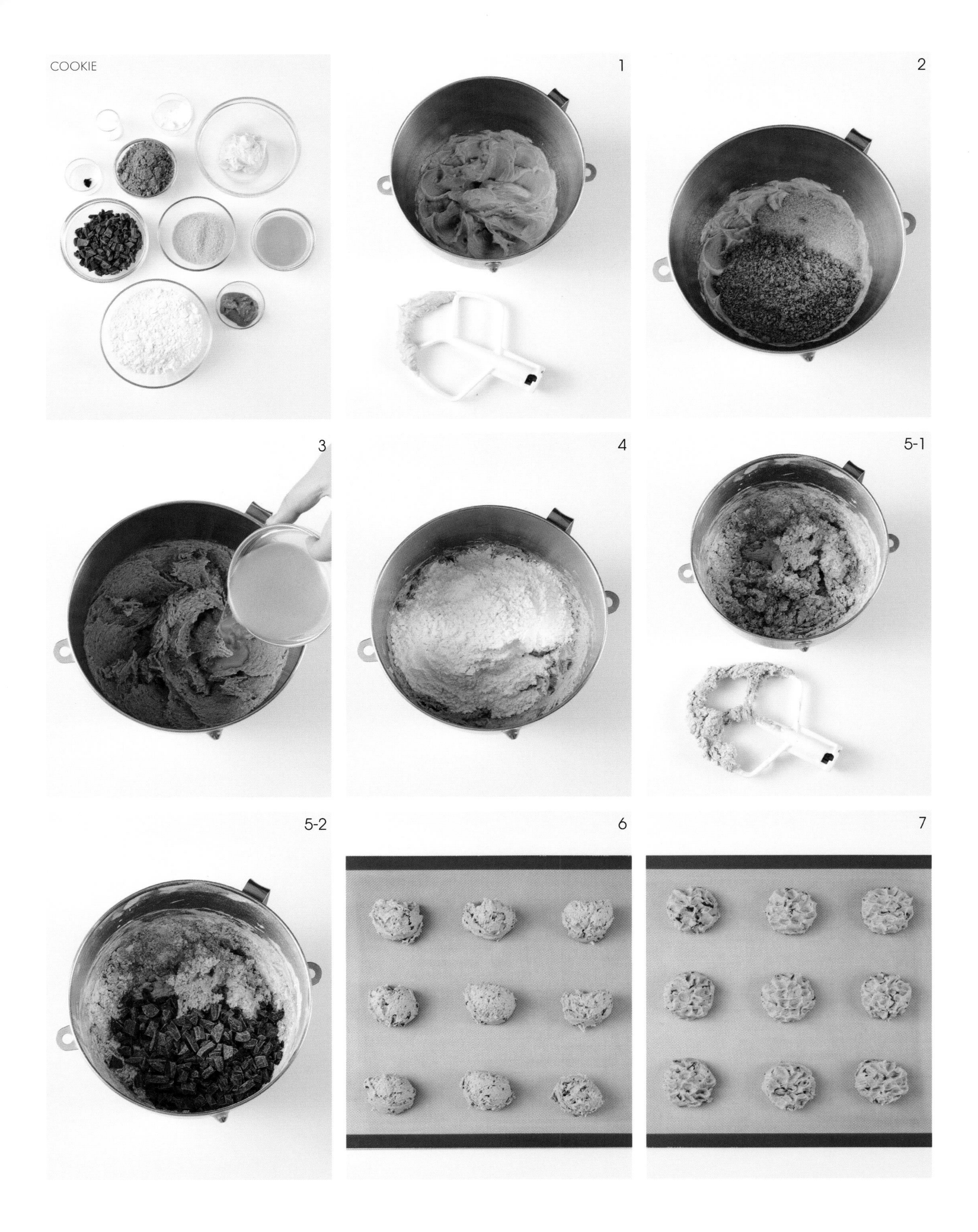
COOKIE
1
2
3
4
5-1
5-2
6
7

Process

쿠키

1. 실온 상태의 버터를 부드럽게 풀어준 후 백된장, 바닐라빈 페이스트를 넣고 혼합한다.
2. 원당, 무스코바도 설탕을 넣고 크림화한다.
3. 실온 상태의 달걀전란을 조금씩 나눠 넣어가며 계속해서 크림화한다.
4. 체 친 박력분, 베이킹파우더, 베이킹소다를 넣고 혼합한다.
5. 80% 정도 혼합되었을 때 조각낸 다크초콜릿을 넣고 완벽하게 혼합한다.
6. 반죽을 80g씩 분할한 후 냉장고에서 최소 2시간 휴지시킨다.
7. 손가락을 이용해 반죽에 자연스러운 질감을 준 후 170℃ 오븐에서 10~11분간 굽는다.

COOKIE

1. Soften room temperature butter. Add white soybean paste and vanilla bean paste to combine.
2. Beat in raw sugar and muscovado sugar to cream the butter mixture.
3. Gradually add room temperature whole eggs little by little while continuing to beat.
4. Add sifted cake flour, baking powder, and baking soda and mix.
5. When the dough is about 80% combined, add chunks of dark chocolate and mix thoroughly.
6. Divide the dough into 80 grams and refrigerate for at least 2 hours.
7. Use fingers to give the dough a natural texture and bake for 10~11 minutes at 170°C.

마무리 8. 오븐에서 구워져 나온 쿠키가 식기 전에 원형 틀로 쿠키 가장자리를 굴려 균일한 사이즈로 모양을 잡아준다.

9. 게랑드소금을 토핑한다.

FINISH

8. Before the cookies cool down, roll the edges of the cookies with a round mold to shape them into uniform size.
9. Sprinkle Guérande salt.

25 DOUBLE CHOCOLATE COOKIE

더블 초콜릿 쿠키

ingredients - 12 cookies

NUTS FREE

쿠키

버터 170g
원당 60g
무스코바도 설탕 110g
게랑드소금 1g
달걀전란 102g
중력분 170g
박력분 85g
카카오파우더 43g
베이킹파우더 2.6g
베이킹소다 1.7g
밀크초콜릿 128g
(JIVARA LATTE 40%)
다크초콜릿 128g
(GUANAJA 70%)

기타

밀크초콜릿
(JIVARA LATTE 40%)
다크초콜릿
(GUANAJA 70%)

NUTS FREE

COOKIE

170g Butter
60g Raw sugar
110g Muscovado sugar
1g Guérande salt
102g Whole eggs
170g All-purpose flour
85g Cake flour
43g Cacao powder
2.6g Baking powder
1.7g Baking soda
128g Milk chocolate
(JIVARA LATTE 40%)
128g Dark chocolate
(GUANAJA 70%)

OTHER

Milk chocolate
(JIVARA LATTE 40%)
Dark chocolate
(GUANAJA 70%)

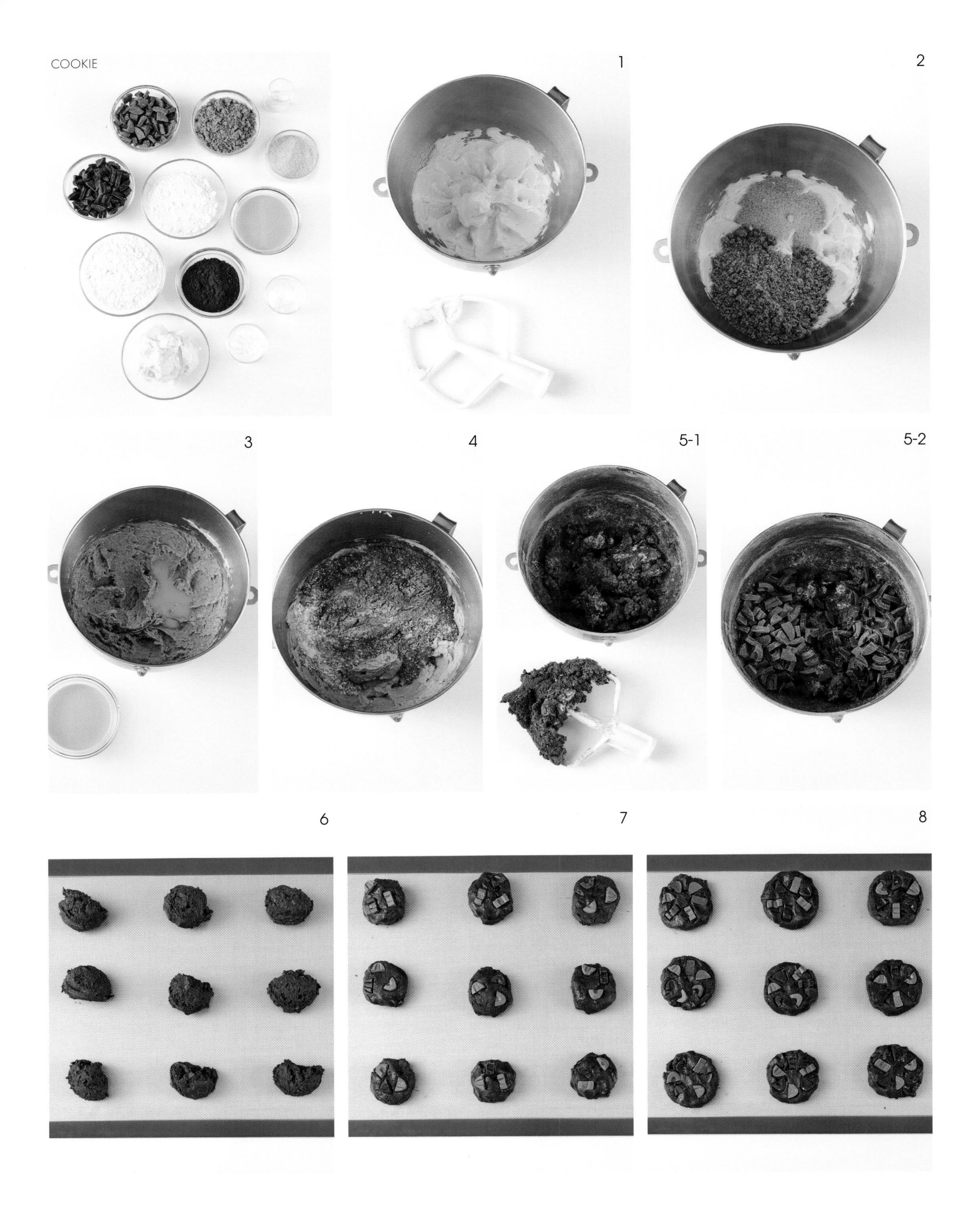
COOKIE
1
2
3
4
5-1
5-2
6
7
8

Process

쿠키

1. 실온 상태의 버터를 부드럽게 풀어준다.
2. 원당, 무스코바도 설탕, 게랑드소금을 넣고 크림화한다.
3. 실온 상태의 달걀전란을 조금씩 나누어 넣어가며 계속해서 크림화한다.
4. 체 친 중력분, 박력분, 카카오파우더, 베이킹파우더, 베이킹소다를 넣고 혼합한다.
5. 80% 정도 혼합되었을 때 조각낸 밀크초콜릿, 다크초콜릿을 넣고 완벽하게 혼합한다.
6. 반죽을 80g씩 분할한 후 냉장고에서 최소 2시간 휴지시킨다.
7. 반죽을 둥글게 성형한 후 조각낸 밀크초콜릿과 다크초콜릿을 토핑한다.
8. 반죽을 약 2cm 높이로 둥글납작하게 눌러 모양을 잡아준 후 180℃ 오븐에서 10~11분간 굽는다.

   * 오븐에서 구워져 나온 쿠키가 식기 전에 원형 틀로 쿠키 가장자리를 굴려 균일한 사이즈로 모양을 잡아준다.

COOKIE

1. Soften the room-temperature butter.
2. Beat in raw sugar, muscovado sugar, and Guérande salt.
3. Gradually add room temperature whole eggs little by little and continue to beat.
4. Add sifted all-purpose flour, cake flour, cacao powder, baking powder, and baking soda, and mix.
5. When the dough is about 80% combined, add chunks of milk chocolate and dark chocolate and mix thoroughly.
6. Divide the dough into 80 grams and refrigerate for at least 2 hours.
7. Shape the dough round and top with milk chocolate and dark chocolate pieces.
8. Slightly flatten the dough to about 2 cm in height, and bake for 10~11 minutes at 180°C.

   * Before the cookies cool down, roll the edges of the cookies with a round mold to shape them into uniform size.

26 VEGAN NUTS COOKIE

비건 넛츠 쿠키

ingredients - 9 cookies

VEGAN

쿠키

통밀가루 141g
중력분 141g
베이킹소다 3.2g
베이킹파우더 3.2g
무스코바도 설탕 134g
게랑드소금 3.2g
다크초콜릿 79g
(CARAIBE 66%)
피칸 39g
아몬드 39g
포도씨유 147g
오트밀크 65g

기타

피칸
다크초콜릿
(CARAIBE 66%)
아몬드

VEGAN

COOKIE

141g Whole wheat flour
141g All-purpose flour
3.2g Baking soda
3.2g Baking powder
134g Muscovado sugar
3.2g Guérande salt
79g Dark chocolate
(CARAIBE 66%)
39g Pecans
39g Almonds
147g Grapeseed oil
65g Oat milk

OTHER

Pecans
Dark chocolate
(CARAIBE 66%)
Almonds

COOKIE
1-1
1-2
2
3-1
3-2
4
5

Process

쿠키

1. 볼에 체 친 통밀가루, 중력분, 베이킹소다, 베이킹파우더에 무스코바도 설탕, 게랑드소금, 조각낸 다크초콜릿, 피칸, 아몬드를 넣고 혼합한다.

   * 피칸은 160℃ 오븐에서 약 15분간 로스팅한 후 4~5등분으로 잘라 준비한다.

   * 아몬드는 140℃ 오븐에서 약 30분간 로스팅한 후 3등분으로 잘라 준비한다.

2. 포도씨유, 오트밀크를 한데 섞어준다.
3. 1에 넣고 혼합한다.
4. 반죽을 85g씩 분할한 후 모양을 잡아준다.
5. 조각낸 피칸, 다크초콜릿, 아몬드를 보기 좋게 토핑한 후 180℃ 오븐에서 10~11분간 굽는다.

   * 오븐에서 구워져 나온 쿠키가 식기 전에 원형 틀로 쿠키 가장자리를 굴려 균일한 사이즈로 모양을 잡아준다.

COOKIE

1. Sift whole wheat flour, all-purpose flour, baking soda, and baking powder, and combine with sugar, muscovado sugar, Guérande salt, chunks of dark chocolate, pecans, and almonds.

   * Roast pecans for about 15 minutes at 160°C and cut them into 4~5 pieces to use.

   * Roast almonds for about 30 minutes at 140°C and cut into 3 pieces to use.

2. Combine grapeseed oil and oat milk in a bowl.
3. Add into 1 and mix.
4. Divide into 85 grams and shape them.
5. Arrange chunks of pecans, dark chocolate, and almonds on top; bake for 10~11 minutes at 180°C.

   * Before the cookies cool down, roll the edges of the cookies with a round mold to shape them into uniform size.

27 EARL GREY COOKIE

얼그레이 쿠키

ingredients - 48 cookies

NUTS FREE

오렌지 얼그레이 토핑 설탕

원당 200g
오렌지제스트 3g
얼그레이파우더 5g

얼그레이 사블레

버터 194g
강력분 63g
박력분 215g
얼그레이파우더 8.1g
게랑드소금 2g
원당 36g
무스코바도 설탕 51g
오렌지제스트 1g
달걀흰자 10g

NUTS FREE

ORANGE EARL GREY SUGAR TOPPING

200g Raw sugar
3g Orange zest
5g Earl grey powder

EARL GREY SABLE

194g Butter
63g Bread flour
215g Cake flour
8.1g Earl grey powder
2g Guérande salt
36g Raw sugar
51g Muscovado sugar
1g Orange zest
10g Egg whites

ORANGE EARL GREY SUGAR TOPPING
1
2
EARL GREY SABLE
3
4
5
6
7
8

Process

오렌지 얼그레이 토핑 설탕

1. 푸드프로세서에 모든 재료를 넣고 믹싱한다.
2. 유산지를 깐 철판에 펼친 후 40℃ 오븐에서 약 2시간 동안 건조시킨 후 사용한다.

얼그레이 사블레

3. 푸드프로세서에 사방 2cm 크기로 자른 버터, 강력분, 박력분을 넣고 사블라주한다.
   * 모든 재료는 차가운 상태로 준비한다.
4. 얼그레이파우더, 게랑드소금, 원당, 무스코바도설탕, 오렌지제스트를 넣고 가볍게 믹싱한다.
5. 달걀흰자를 넣고 가볍게 믹싱한다.
6. 푸드프로세서에서 꺼낸 반죽을 스크래퍼로 이겨 매끄럽게 혼합한다.
7. 7mm 두께로 밀어 편 후 냉장고에서 약 2시간 휴지시킨다.
8. 35 × 35mm 크기로 재단한다.

ORANGE EARL GREY SUGAR TOPPING

1. Combine all the ingredients in a food processor.
2. Spread over a baking tray lined with parchment paper and dry for about 2 hours at 40°C to use.

EARL GREY SABLE

3. Sablage butter cut into 2 cm cubes, bread flour, and cake flour in a food processor.
   * Prepare all the ingredients cold.
4. Add earl grey powder, Guérande salt, raw sugar, muscovado sugar, and orange zest; mix briefly.
5. Add egg whites and mix briefly.
6. Remove the dough from the food processor. Knead it with a scraper to make a smooth mixture.
7. Roll out to 7 mm and refrigerate for about 2 hours.
8. Cut to 35 × 35 mm.

마무리

9. 재단면에 오렌지 얼그레이 토핑 설탕을 묻혀준다.
10. 에어 매트에 팬닝한 후 165℃ 오븐에서 20~25분간 굽는다.

FINISH

9. Coat the sides with earl grey sugar topping.
10. Place over a perforated mat and bake for 20~25 minutes at 165°C.

28 LEMON & BASIL COOKIE

레몬 & 바질 쿠키

ingredients - 48 cookies

EGG FREE　NUTS FREE

레몬 & 바질 토핑 설탕

원당 200g
레몬제스트 3g
바질 5g

레몬 & 바질 사블레

버터 191g
강력분 62g
박력분 212g
원당 85g
레몬제스트 7g
게랑드소금 2g
바질 21g

EGG FREE　NUTS FREE

LEMON & BASIL SUGAR TOPPING

200g Raw sugar
3g Lemon zest
5g Basil

LEMON & BASIL SABLE

191g Butter
62g Bread flour
212g Cake flour
85g Raw sugar
7g Lemon zest
2g Guérande salt
21g Basil

LEMON & BASIL SUGAR TOPPING
1
2
LEMON & BASIL SABLE
3
4
5
6
7

Process

레몬 & 바질 토핑 설탕

1. 푸드프로세서에 모든 재료를 넣고 믹싱한다.
2. 유산지를 깐 철판에 펼친 후 40℃ 오븐에서 약 2시간 동안 건조시킨 후 사용한다.

레몬 & 바질 사블레

3. 푸드프로세서에 사방 2cm 크기로 자른 버터, 강력분, 박력분을 넣고 사블라주한다.
   * 모든 재료는 차가운 상태로 준비한다.
4. 원당, 레몬제스트, 게랑드소금, 바질을 넣고 가볍게 믹싱한다.
5. 푸드프로세서에서 꺼낸 반죽을 스크래퍼로 이겨 매끄럽게 혼합한다.
6. 7mm 두께로 밀어 편 후 냉장고에서 약 2시간 휴지시킨다.
7. 35 × 35mm 크기로 재단한다.

LEMON & BASIL SUGAR TOPPING

1. Combine all the ingredients in a food processor.
2. Spread over a baking tray lined with parchment paper and dry for about 2 hours at 40°C to use.

LEMON & BASIL SABLE

3. Sablage butter cut into 2 cm cubes, bread flour, and cake flour in a food processor.
   * Prepare all the ingredients cold.
4. Add raw sugar, lemon zest, Guérande salt, and basil; mix briefly.
5. Remove the dough from the food processor. Knead it with a scraper to make a smooth mixture.
6. Roll out to 7 mm and refrigerate for about 2 hours.
7. Cut to 35 × 35 mm.

마무리

8. 재단면에 레몬 바질 토핑 설탕을 묻혀준다.
9. 에어 매트에 팬닝한 후 165℃ 오븐에서 20~25분간 굽는다.

FINISH	8.	Coat the sides with lemon basil sugar topping.
	9.	Place over a perforated mat and bake for 20~25 minutes at 165°C.

29 SEAWEED COOKIE

감태 쿠키

ingredients - 48 cookies

NUTS
FREE

쿠키

버터 186g
강력분 62g
박력분 207g
김 9g
감태 9g
게랑드소금 2g
유기농 사탕수수당 83g
(잘레스 이타자 유기농 황설탕)
달걀흰자 21g

기타

감태가루

NUTS
FREE

COOKIE

186g Butter
62g Bread flour
207g Cake flour
9g Seaweed
9g Gamtae
2g Guérande salt
83g Organic cane sugar
(Jalles Itaja Organic Sugar)
21g Egg whites

OTHER

Gamtae powder

COOKIE
1-1
1-2
2
3
4
5
FINISH
6
7

Process

쿠키

1. 푸드프로세서에 사방 2cm 크기로 자른 버터, 강력분, 박력분을 넣고 사블라주한다.
   * 모든 재료는 차가운 상태로 준비한다.
2. 김, 감태, 게랑드소금, 유기농 사탕수수당을 넣고 가볍게 믹싱한다.
3. 달걀흰자를 넣고 믹싱한다.
4. 푸드프로세서에서 꺼낸 반죽을 스크래퍼로 이겨 매끄럽게 혼합한다.
5. 7mm 두께로 밀어 편 후 냉장고에서 약 2시간 휴지시킨다.

마무리

6. 35 × 35mm 크기로 재단한 후 재단면에 감태가루를 묻혀준다.
7. 에어 매트에 팬닝한 후 165℃ 오븐에서 20~25분간 굽는다.

COOKIE

1. Sablage butter cut into 2 cm cubes, bread flour, and cake flour in a food processor.
   * Prepare all the ingredients cold.
2. Add seaweed, gamtae, Guérande salt, and organic cane sugar; mix briefly.
3. Add egg whites and mix.
4. Remove the dough from the food processor. Knead it with a scraper to make a smooth mixture.
5. Roll out to 7 mm and refrigerate for about 2 hours.

FINISH

6. Cut to 35 × 35 mm and coat the sides with gamtae powder.
7. Place over a perforated mat and bake for 20~25 minutes at 165°C.

30 CINNAMON & PECAN COOKIE

시나몬 & 피칸 쿠키

ingredients - 48 cookies

EGG FREE

캐러멜라이즈 피칸*

시나몬파우더 1.5g
설탕A 37g
물 28g
설탕B 121g
피칸 213g

시나몬 토핑 슈거

원당 200g
시나몬파우더 3g

시나몬 & 피칸 쿠키

버터 152g
원당 67g
게랑드소금 5g
강력분 51g
박력분 168g
시나몬파우더 2.5g
캐러멜라이즈 피칸* 135g

EGG FREE

CARAMELIZED PECANS*

1.5g Cinnamon powder
37g Sugar A
28g Water
121g Sugar B
213g Pecans

CINNAMON SUGAR TOPPING

200g Raw sugar
3g Cinnamon powder

CINNAMON & PECAN COOKIE

152g Butter
67g Raw sugar
5g Guérande salt
51g Bread flour
168g Cake flour
2.5g Cinnamon powder
135g Caramelized pecans*

CARAMELIZED PECANS
1
2
3
4
5
6
7
8

Process

캐러멜라이즈 피칸

1. 피칸은 160℃ 오븐에서 8분간 구워 준비한다.
2. 시나몬파우더, 설탕A를 혼합해 준비한다.
3. 냄비에 물, 설탕B를 넣고 가열한다.
4. 118℃가 되면 따뜻한 상태의 피칸을 넣고 섞어준다.
5. 시럽이 하얗게 재결정화 상태가 될 때까지 계속해서 저어준다.
6. 다시 불에 올려 골고루 저어주며 진한 갈색이 날 때까지 계속해서 가열한다.
7. 불에서 내려 **2**를 넣고 버무린 후 피칸이 서로 달라붙지 않고 한 알씩 잘 떨어지도록 한다.
8. 실팻에 펼쳐 완전히 식힌 후 조각내 사용한다.

CARAMELIZED PECANS

1. Roast pecans for 8 minutes at 160°C.
2. Combine cinnamon powder and sugar A; set aside.
3. Heat water and sugar B in a pot.
4. When it reaches 118°C, add warm pecans and mix.
5. Continue to stir until the syrup recrystallizes and turns white.
6. Put it back over the heat and continue stirring until it turns dark brown.
7. Remove from the heat and mix with **2**, while taking care that the nuts will separate easily from each other.
8. Spread over a Silpat, cool completely, and coarsely chop to use.

CINNAMON SUGAR TOPPING
9
CINNAMON & PECAN COOKIE
10
11
12
13
FINISH
14
15

시나몬 토핑 슈거

9. 푸드프로세서에 원당, 시나몬파우더를 넣고 믹싱한다.

시나몬 피칸 쿠키

10. 믹싱볼에 실온 상태의 버터, 원당, 게랑드소금을 넣고 믹싱한다.
11. 체 친 강력분, 박력분, 시나몬파우더를 넣고 혼합한다.
12. 80% 정도 혼합되었을 때 조각낸 캐러멜라이즈 피칸을 넣고 완벽하게 혼합한다.
13. 두께 7mm로 밀어 편 후 냉장고에서 약 2시간 휴지시킨다.

마무리

14. 32×32mm 크기로 재단한 후 재단면에 시나몬 토핑 슈거를 묻힌다.
15. 에어 매트에 팬닝한 후 160℃ 오븐에서 20~24분간 굽는다.

CINNAMON SUGAR TOPPING

9. Combine raw sugar and cinnamon powder in a food processor.

CINNAMON & PECAN COOKIE

10. Mix room-temperature butter, raw sugar, and Guérande salt in a mixing bowl.
11. Mix with sifted bread flour, cake flour, and cinnamon powder.
12. When the dough is about 80% combined, add chopped caramelized pecan pieces; mix thoroughly.
13. Roll out to 7 mm and refrigerate for about 2 hours.

FINISH

14. Cut to 32 × 32 mm and coat the sides with cinnamon sugar topping.
15. Place over a perforated mat and bake for 20~24 minutes at 160°C.

31 VEGAN COCO

비건 코코

ingredients - 48 cookies

비건 코코

코코넛슈거 142g
거친 쌀가루 119g
카카오파우더 17g
아몬드파우더 145g
게랑드소금 2.2g
두유 43g
코코넛오일 113g

기타

코코넛 슈거

VEGAN COCO

142g Coconut sugar
119g Coarse rice flour
17g Cacao powder
145g Almond powder
2.2g Guérande salt
43g Soy milk
113g Coconut oil

OTHER

Coconut sugar

1
2-1
2-2
2-3
3
FINISH
4
5

Process

비건 코코

1. 볼에 코코넛슈거, 거친 쌀가루, 카카오파우더, 아몬드파우더, 게랑드소금을 넣고 혼합한다.
2. 두유와 녹인 코코넛오일을 넣고 반죽을 한 덩어리로 만든다.
3. 7mm 두께로 밀어 편 후 냉장고에서 약 2시간 휴지시킨다.

마무리

4. 35 × 35mm 크기로 재단한 후 재단면에 코코넛슈거를 묻혀준다.
5. 에어 매트에 팬닝한 후 165℃ 오븐에서 20~25분간 굽는다.

VEGAN COCO

1. Combine coconut sugar, coarse rice flour, cacao powder, almond powder, and Guérande salt in a bowl.
2. Add soy milk and melted coconut oil; mix until the dough comes together.
3. Roll out to 7 mm and refrigerate for about 2 hours.

FINISH

4. Cut to 35 × 35 mm and coat the sides with coconut sugar.
5. Place over a perforated mat and bake for 20~25 minutes at 165°C.

32 BERRY BERRY MONAKA FLORENTINE

베리베리 모나카 플로랑탱

ingredients - 32 cookies

EGG FREE

베리베리 모나카 플로랑탱

원당 77g
멸균 생크림 54g
꿀 25g
물엿 25g
버터 25g
백아몬드 슬라이스 147g
전처리한 건조 크랜베리 46g

기타

모나카 셸
라즈베리 크리스피
(SOSA *wetproof*)

EGG FREE

BERRY BERRY MONAKA FLORENTINE

77g Raw sugar
54g UHT heavy cream
25g Honey
25g Corn syrup
25g Butter
147g Blanched almond slices
46g Pre-treated dried cranberries

OTHER

Monaka shells
Raspberry crispy
(SOSA *wetproof*)

PRE-TREATED CRANBERRIES
1
2
BERRY BERRY MONAKA FLORENTINE
3
4
5

Process

크랜베리 전처리

1. 끓는 물에 건조 크랜베리를 데친다.
2. 물기를 충분히 제거한 후 잘게 조각낸다.

베리베리 모나카 플로랑탱

3. 냄비에 원당, 멸균 생크림, 꿀, 물엿, 버터를 넣고 원당과 버터가 완전히 녹을 때까지 가열한다.
4. 백아몬드 슬라이스, 전처리한 건조 크랜베리를 넣고 버무린다.
5. 냉장고에서 약 90분간 휴지시킨다.

PRE-TREATED CRANBERRIES

1. Blanch dried cranberries in boiling water.
2. Drain water sufficiently and chop finely.

BERRY BERRY MONAKA FLORENTINE

3. Heat raw sugar, UHT heavy cream, honey, corn syrup, and butter in a pot until the raw sugar and butter melt.
4. Add blanched almond slices and pre-treated dried cranberries; mix thoroughly.
5. Refrigerate for about 90 minutes.

마무리

6. 모나카 셸에 12g씩 채워준 후 고르게 펼친다.
7. 165℃ 오븐에서 17~18분간 굽는다.
8. 오븐에서 나온 직후 라즈베리 크리스피를 토핑한다.

FINISH

6. Put 12 grams in the monaka shells and spread evenly.
7. Bake for 17~18 minutes at 165°C.
8. Sprinkle with raspberry crispy immediately after baking.

33 SALTED MONAKA FLORENTINE

솔티드 모나카 플로랑탱

ingredients - 32 cookies

EGG FREE

솔티드 모나카 플로랑탱

원당 82g
멸균 생크림 45g
꿀 27g
물엿 27g
버터 27g
백아몬드 슬라이스 193g

기타

모나카 셸
게랑드소금

EGG FREE

SALTED MONAKA FLORENTINE

82g Raw sugar
45g UHT heavy cream
27g Honey
27g Corn syrup
27g Butter
193g Blanched almond slices

OTHER

Monaka shells
Guérande salt

SALTED MONAKA FLORENTINE
1
2
3
FINISH
4-1
4-2
5
6

Process

솔티드 모나카 플로랑탱

1. 냄비에 원당, 멸균 생크림, 꿀, 물엿, 버터를 넣고 원당과 버터가 완전히 녹을 때까지 가열한다.
2. 백아몬드 슬라이스를 넣고 버무린다.
3. 냉장고에서 약 90분간 휴지시킨다.

마무리

4. 모나카 셸에 12g씩 채워준 후 고르게 펼친다.
5. 165℃ 오븐에서 17~18분간 굽는다.
6. 구워져 나온 직후 게랑드소금을 토핑한다.

SALTED MONAKA FLORENTINE

1. Heat raw sugar, UHT heavy cream, honey, corn syrup, and butter in a pot until the raw sugar and butter melt.
2. Add blanched almond slices and mix thoroughly.
3. Refrigerate for about 90 minutes.

FINISH

4. Put 12 grams in the monaka shells and spread evenly.
5. Bake for 17~18 minutes at 165°C.
6. Sprinkle with Guérande salt immediately after baking.

[34] SESAME MONAKA FLORENTINE

깨 모나카 플로랑탱

ingredients - 32ea

깨 모나카 플로랑탱

원당 73g
멸균 생크림 49g
꿀 24g
물엿 24g
버터 24g
흑임자 56g
참깨 56g
아몬드 분태 93g
레몬제스트 0.7g

기타

모나카 셸

SESAME MONAKA FLORENTINE

73g Raw sugar
49g UHT heavy cream
24g Honey
24g Corn syrup
24g Butter
56g Black sesame seeds
56g Sesame seeds
93g Chopped almonds
0.7g Lemon zest

OTHER

Monaka shells

SESAME MONAKA FLORENTINE
1
2
3
4
FINISH
5
6

Process

깨 모나카 플로랑탱

1. 냄비에 원당, 멸균 생크림, 꿀, 물엿, 버터를 넣고 원당과 버터가 완전히 녹을 때까지 가열한다.
2. 흑임자, 참깨, 아몬드 분태를 넣고 버무린다.
3. 레몬제스트를 넣고 혼합한다.
4. 냉장고에서 약 90분간 휴지시킨다.

마무리

5. 모나카 셸에 11g씩 채워준 후 고르게 펼친다.
6. 165℃ 오븐에서 17~18분간 굽는다.

SESAME MONAKA FLORENTINE

1. Heat raw sugar, UHT heavy cream, honey, corn syrup, and butter in a pot until the raw sugar and butter melt.
2. Add black sesame seeds, sesame seeds, and chopped almonds; mix thoroughly.
3. Add lemon zest and mix.
4. Refrigerate for about 90 minutes.

FINISH

5. Put 11 grams in the monaka shells and spread evenly.
6. Bake for 17~18 minutes at 165°C.

35 COCONUT MONAKA FLORENTINE

코코넛 모나카 플로랑탱

ingredients - 32 cookies

코코넛 모나카 플로랑탱

원당 90g
멸균 생크림 60g
꿀 30g
물엿 30g
버터 30g
칼아몬드 62g
롱코코넛 100g

기타

모나카 셸
게랑드소금

COCONUT MONAKA FLORENTINE

90g Raw sugar
60g UHT heavy cream
30g Honey
30g Corn syrup
30g Butter
62g Slivered almonds
100g Shredded coconuts

OTHER

Monaka shells
Guérande salt

1
2
3
FINISH
4
5

Process

코코넛 모나카 플로랑탱

1. 냄비에 원당, 멸균 생크림, 꿀, 물엿, 버터를 넣고 원당과 버터가 완전히 녹을 때까지 가열한다.
2. 구운 칼아몬드, 구운 롱코코넛을 넣고 버무린다.
   * 칼아몬드와 롱코코넛은 160℃ 오븐에서 약 8분간 구워 준비한다.
3. 냉장고에서 약 90분간 휴지시킨다.

마무리

4. 모나카 셸에 11g씩 채워준 후 고르게 펼친다. 165℃ 오븐에서 17~18분간 굽는다.
5. 구워져 나온 직후 게랑드소금을 토핑한다.

COCONUT MONAKA FLORENTINE

1. Heat raw sugar, UHT heavy cream, honey, corn syrup, and butter in a pot until the raw sugar and butter melt.
2. Add roasted slivered almonds and shredded coconuts; mix thoroughly.
   * Roast slivered almonds and shredded coconuts for 8 minutes at 160°C.
3. Refrigerate for about 90 minutes.

FINISH

4. Put 11 grams in the monaka shells and spread evenly, and bake for 17~18 minutes at 165°C.
5. Sprinkle with Guérande salt immediately after baking.

RASPBERRY

36 RASPBERRY SAND COOKIE

라즈베리 샌드 쿠키

ingredients - Ø 6.5cm, 12 cookies

GLUTEN FREE

쿠키

슈거파우더 88g
게랑드소금 2.5g
아몬드파우더 88g
고운 쌀가루 142g
베이킹파우더 0.6g
붉은색 천연 식용 색소 11.3g
버터 108g
달걀흰자 11g

라즈베리 버터 가나슈

버터 57g
전화당 13g
프람보아즈초콜릿 167g
(FRAMBOISE INSPIRATION)
프람보아즈 리큐어 13g
(DIJON FRAMBOISE)

라즈베리 충전물

NH펙틴 3g
설탕 37g
냉동 라즈베리 88g
라즈베리 퓌레 88g
레몬주스 30g
전화당 19g
트레할로스 27g
프람보아즈 리큐어 9g
(DIJON FRAMBOISE)

초콜릿 벨벳 스프레이

카카오버터 100g
프람보아즈초콜릿 100g
(FRAMBOISE INSPIRATION)
포도씨유 20g
붉은색 천연 식용 색소

GLUTEN FREE

COOKIE

88g Powdered sugar
2.5g Guérande salt
88g Almond powder
142g Fine rice powder
0.6g Baking powder
11.3g Red natural food color
108g Butter
11g Egg whites

RASPBERRY BUTTER GANACH

57g Butter
13g Inverted sugar
167g Framboise chocolate
(FRAMBOISE INSPIRATION)
13g Framboise liqueur

RASPBERRY FILLING

3g Pectin NH
37g Sugar
88g Frozen raspberries
88g Raspberry puree
30g Lemon juice
19g Inverted sugar
27g Trehalose
9g Framboise liqueur
(DIJON FRAMBOISE)

CHOCOLATE VELVET SPRAY

100g Cacao butter
100g Framboise chocolate
(FRAMBOISE INSPIRATION)
20g Grapeseed oil
Red natural food color

COOKIE
1
2-1
2-2
3
4
5
6
7

Process

쿠키

1. 푸드프로세서에 버터와 달걀흰자를 제외한 모든 재료를 넣고 혼합한다.
2. 버터, 달걀흰자를 넣고 반죽이 뭉치기 시작할 때까지 혼합한다.
   * 버터는 20℃ 정도의 상태로 준비한다.
3. 푸드프로세서에서 꺼낸 반죽을 스크래퍼로 이겨 매끄럽게 혼합한다.
4. 3mm 두께로 밀어 편 후 냉동고에서 단단하게 굳힌다.
5. 6.5cm 국화 모양 커터로 커팅한 후 두 장의 에어 매트 사이에 팬닝한다.
6. 160℃ 오븐에서 15분간 굽는다.
7. 구워져 나온 직후 녹인 카카오버터(분량 외)를 발라준다.

COOKIE

1. Combine all the ingredients, except butter and egg whites, in a food processor.
2. Add butter and egg whites and mix until the dough starts to come together.
   * Prepare butter to about 20°C.
3. Remove the dough from the food processor. Knead it with a scraper to make a smooth mixture.
4. Roll out to 3 mm and set it cold in a freezer.
5. Cut out with a 6.5 cm round crinkled cutter and place them between two perforated mats.
6. Bake for 15 minutes at 160°C.
7. Brush with melted cacao butter (other than requested) immediately after baking.

RASPBERRY BUTTER GANACHE
8
9
10
RASPBERRY FILLING
11
12
13
14
15
16

라즈베리 버터 가나슈

8. 볼에 포마드 상태의 버터, 전화당을 넣고 부드럽게 풀어준다.
9. 녹인 프람보아즈초콜릿(30℃)을 조금씩 나눠 넣어가며 혼합한다.
10. 프람보아즈 리큐어를 넣고 혼합한다.

라즈베리 충전물

11. NH펙틴과 설탕을 섞어준다.
12. 볼에 **11**과 냉동 라즈베리, 라즈베리 퓌레, 레몬주스, 전화당, 트레할로스를 넣고 섞어준다.
13. 냉동 라즈베리가 충분히 해동되고 과즙이 빠져나올 때까지 실온에 12시간 정도 둔다.
14. **13**을 냄비에 옮겨 45Brix가 될 때까지 가열한다.
15. 불에서 내린 후 프람보아즈 리큐어를 넣고 섞어준다.
16. 트레이에 부어 차갑게 식힌 후 부드럽게 풀어 사용한다.

RASPBERRY BUTTER GANACHE

8. Soften room-temperature butter and inverted sugar in a bowl.
9. Gradually add melted framboise chocolate (30°C) little by little to combine.
10. Add framboise liqueur and mix.

RASPBERRY FILLING

11. Combine pectin NH and sugar.
12. Mix **11** with frozen raspberries, raspberry puree, lemon juice, inverted sugar, and trehalose in a bowl.
13. Let stand at an ambient temperature for about 12 hours to defrost frozen raspberries sufficiently and extract the juice.
14. Heat **13** in a pot until it reaches 45 Brix.
15. Remove from heat and mix with framboise liqueur.
16. Pour onto a tray, cool sufficiently, and soften to use.

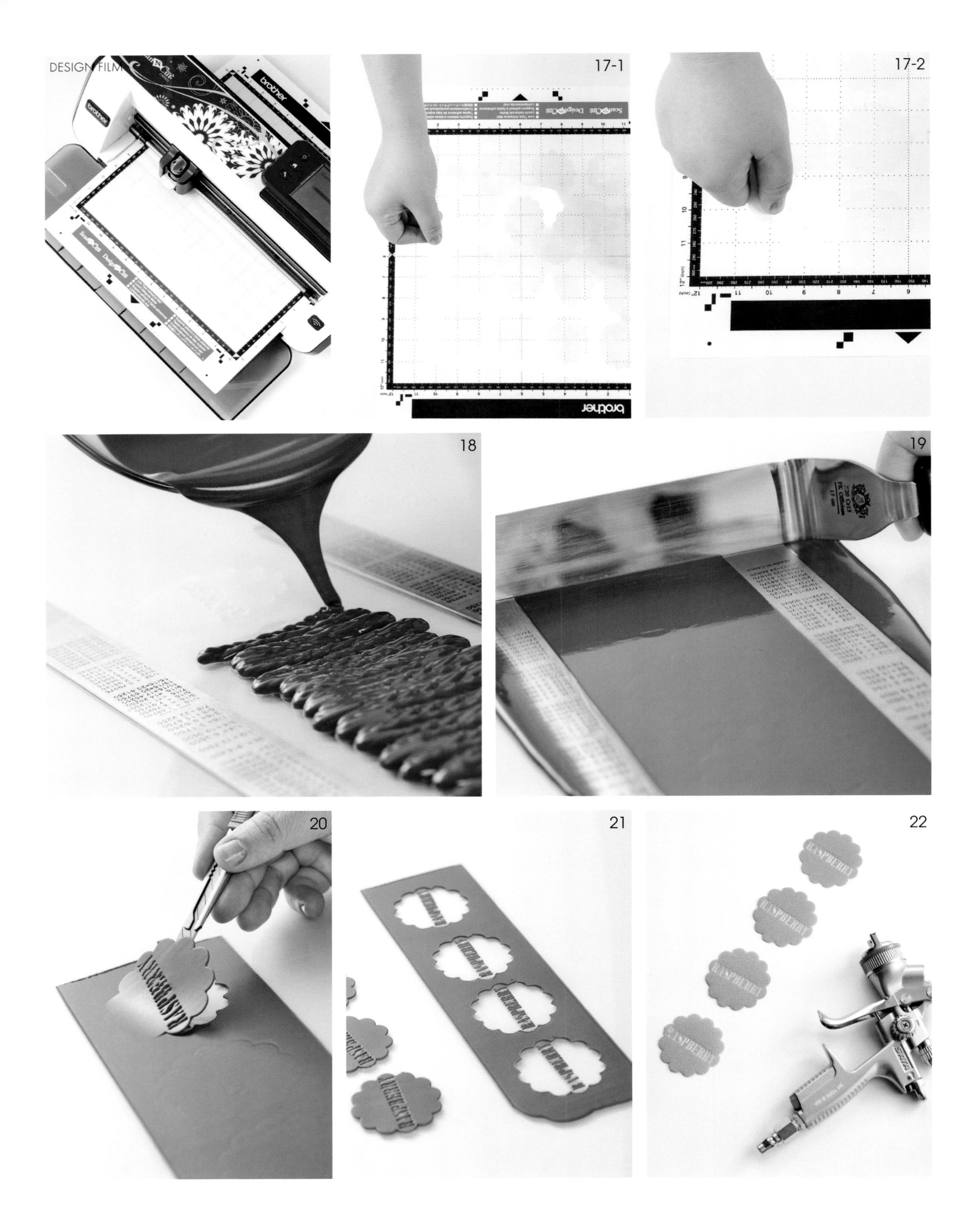

17-1
17-2
18
19
20
21
22
brother
RASPBERRY

디자인 필름

17. 커팅기를 이용해 디자인한 도안을 출력한다.
18. 커팅한 필름 위에 템퍼링한 프람보아즈초콜릿(분량 외)을 적당량 부어준다.
19. 스패출러를 이용해 초콜릿을 1mm 두께로 얇게 펴준다.
    * 이때 메탈 또는 실리콘 재질의 두께 바를 이용하면 균일한 두께로 밀어 펼 수 있다.
20. 초콜릿이 굳기 시작하면 조심스럽게 필름을 들어올린다.
21. 충분히 수축시킨 후 필름을 떼어내고 냉동고에 30분 동안 넣어둔다.
22. 카카오버터, 프람보아즈초콜릿, 포도씨유, 붉은색 천연 식용 색소를 혼합한 후 차가워진 초콜릿 표면에 분사해 벨벳 느낌을 표현한다.

DESIGN FILM

17. Cut out the desired design using a cutting machine.
18. Pour a moderate amount of tempered framboise chocolate (other than requested) over the cut film.
19. Spread the chocolate thin to 1 mm thickness with a spatula.
    * You can spread to an even thickness when metal or silicone confectionery bars are used.
20. Carefully lift the film when the chocolate starts to set.
21. Let crystallize sufficiently, remove the film, and freeze for 30 minutes.
22. Combine cacao butter, Framboise chocolate, grapeseed oil, and red natural food color; spray on the surface of the cold chocolate to give a velvet finish.

마무리

23. 완전히 식힌 두 장의 쿠키 사이에 라즈베리 버터 가나슈를 파이핑한다.
24. 라즈베리 버터 가나슈 중심부에 라즈베리 충전물을 파이핑한다.
25. 샌딩한 후 수평이 되도록 가볍게 눌러준다.
26. 여분의 초콜릿을 소량 파이핑한다.
27. 장식물을 올려 고정시킨다.

FINISH

23. Pipe raspberry butter ganache between two pieces of cooled cookies.
24. Pipe raspberry filling in the center of the raspberry butter ganache.
25. Press lightly to level the cookies.
26. Pipe a small amount of chocolate on the cookie.
27. Attach the decoration.

SPICE
SPICE
SPICE

37 SPICED SAND COOKIE

스파이스 샌드 쿠키

ingredients - Ø 6.5cm, 12 cookies

GLUTEN FREE

쿠키

슈거파우더 90g
게랑드소금 2.6g
아몬드파우더 90g
고운 쌀가루 132g
베이킹파우더 0.6g
시나몬파우더 6.4g
생강가루 1.5g
정향가루 2.6g
넛맥가루 1.2g
버터 111g
달걀흰자 12g

스파이스 버터 가나슈

버터 53g
전화당 9g
당밀 14g
넛맥가루 0.5g
시나몬파우더 0.9g
통카 0.5g
밀크초콜릿 153g
(JIVARA LATTE 40%)
아니세트 리큐어 20g
(DIJON ANISETTE)

초콜릿 벨벳 스프레이

카카오버터 100g
밀크초콜릿 100g
(JIVARA LATTE 40%)
포도씨유 20g

GLUTEN FREE

COOKIE

90g Powdered sugar
2.6g Guérande salt
90g Almond powder
132g Fine rice powder
0.6g Baking powder
6.4g Cinnamon powder
1.5g Ginger power
2.6g Cloves powder
1.2g Nutmeg powder
111g Butter
12g Egg whites

SPICED BUTTER GANACHE

53g Butter
9g Inverted sugar
14g Molasses
0.5g Nutmeg powder
0.9g Cinnamon powder
0.5g Tonka bean
153g Milk chocolate
(JIVARA LATTE 40%)
20g Anisette liqueur
(DIJON ANISETTE)

CHOCOLATE VELVET SPRAY

100g Cacao butter
100g Milk chocolate
(JIVARA LATTE 40%)
20g Grapeseed oil

COOKIE
1
2-1
2-2
3
4
5
6
7

Process

쿠키

1. 푸드프로세서에 버터와 달걀흰자를 제외한 모든 재료를 넣고 혼합한다.
2. 버터, 달걀흰자를 넣고 반죽이 뭉치기 시작할 때까지 혼합한다.
   * 버터는 20℃ 정도의 상태로 준비한다.
3. 푸드프로세서에서 꺼낸 반죽을 스크래퍼로 이겨 매끄럽게 혼합한다.
4. 3mm 두께로 밀어 편 후 냉동고에서 단단하게 굳힌다.
5. 6.5cm 국화 모양 커터로 커팅한 후 두 장의 에어 매트 사이에 팬닝한다.
6. 160℃ 오븐에서 15분간 굽는다.
7. 구워져 나온 직후 녹인 카카오버터(분량 외)를 발라준다.

COOKIE

1. Combine all the ingredients, except butter and egg whites, in a food processor.
2. Add butter and egg whites and mix until the dough starts to come together.
   * Prepare butter to about 20°C.
3. Remove the dough from the food processor. Knead it with a scraper to make a smooth mixture.
4. Roll out to 3 mm and set it cold in a freezer.
5. Cut out with a 6.5 cm round crinkled cutter and place them between two perforated mats.
6. Bake for 15 minutes at 160°C.
7. Brush with melted cacao butter (other than requested) immediately after baking.

SPICED BUTTER GANACHE
8
9
10
11
FINISH
12
SPICE
13
14
15
16

스파이스 버터 가나슈

8. 볼에 포마드 상태의 버터, 전화당을 넣고 부드럽게 풀어준다.
9. 당밀, 넛맥가루, 시나몬파우더, 곱게 간 통카를 넣고 혼합한다.
10. 녹인 밀크초콜릿(30℃)을 조금씩 나눠 넣어가며 혼합한다.
11. 아니세트 리큐어를 넣고 혼합한다.

마무리

12. 314-315p와 동일한 방법으로 초콜릿 장식물을 준비한다.
13. 완전히 식힌 두 장의 쿠키 사이에 스파이스 버터 가나슈를 파이핑한다.
14. 샌딩한 후 수평이 되도록 가볍게 눌러준다.
15. 여분의 초콜릿을 소량 파이핑한다.
16. 장식물을 올려 고정시킨다.

SPICED BUTTER GANACHE

8. Soften room-temperature butter and inverted sugar in a bowl.
9. Mix with molasses, nutmeg powder, cinnamon powder, and finely grated tonka bean.
10. Gradually add melted milk chocolate (30°C) little by little to combine.
11. Add anisette liqueur and mix.

FINISH

12. Prepare the chocolate decorations the same on pages 314-315.
13. Pipe spiced butter ganache between two pieces of cooled cookies.
14. Press lightly to level the cookies.
15. Pipe a small amount of chocolate on the cookie.
16. Attach the decoration.

PISTACHIO
PISTACHIO

38 PISTACHIO SAND COOKIE

피스타치오 샌드 쿠키

ingredients - Ø 6.5cm, 12 cookies

쿠키

슈거파우더 90g
게랑드소금 2.6g
피스타치오가루 90g
고운 쌀가루 142g
베이킹파우더 0.6g
녹차가루 3g
노란색 천연 식용 색소 0.9g
버터 110g
달걀흰자 12g

피스타치오 버터 가나슈

버터 32g
전화당 12g
피스타치오 페이스트 27g
구운 피스타치오 분태 11g
화이트초콜릿 156g (IVOIRE 35%)
오렌지 리큐어 12g (GRAND MARNIER)

오렌지 충전물

전처리한 오렌지 165g
NH펙틴 8.3g
설탕 58g
오렌지주스 79g
패션푸르트 퓌레 89g
레몬주스 27g
꿀 31g
트레할로스 40g

초콜릿 벨벳 스프레이

카카오버터 100g
화이트초콜릿 100g (IVOIRE 35%)
초록색 천연 식용 색소

COOKIE

90g Powdered sugar
2.6g Guérande salt
90g Pistachio powder
142g Fine rice flour
0.6g Baking powder
3g Matcha powder
0.9g Yellow natural food color
110g Butter
12g Egg whites

PISTACHIO BUTTER GANACHE

32g Butter
12g Inverted sugar
27g Pistachio paste
11g Chopped roasted pistachios
156g White chocolate (IVOIRE 35%)
12g Orange liqueur (GRAND MARNIER)

ORANGE FILLING

165g Pre-treated orange
8.3g Pectin NH
58g Sugar
79g Orange juice
89g Passion fruit puree
27g Lemon juice
31g Honey
40g Trehalose

CHOCOLATE VELVET SPRAY

100g Cacao butter
100g White chocolate (IVOIRE 35%)
Green natural food color

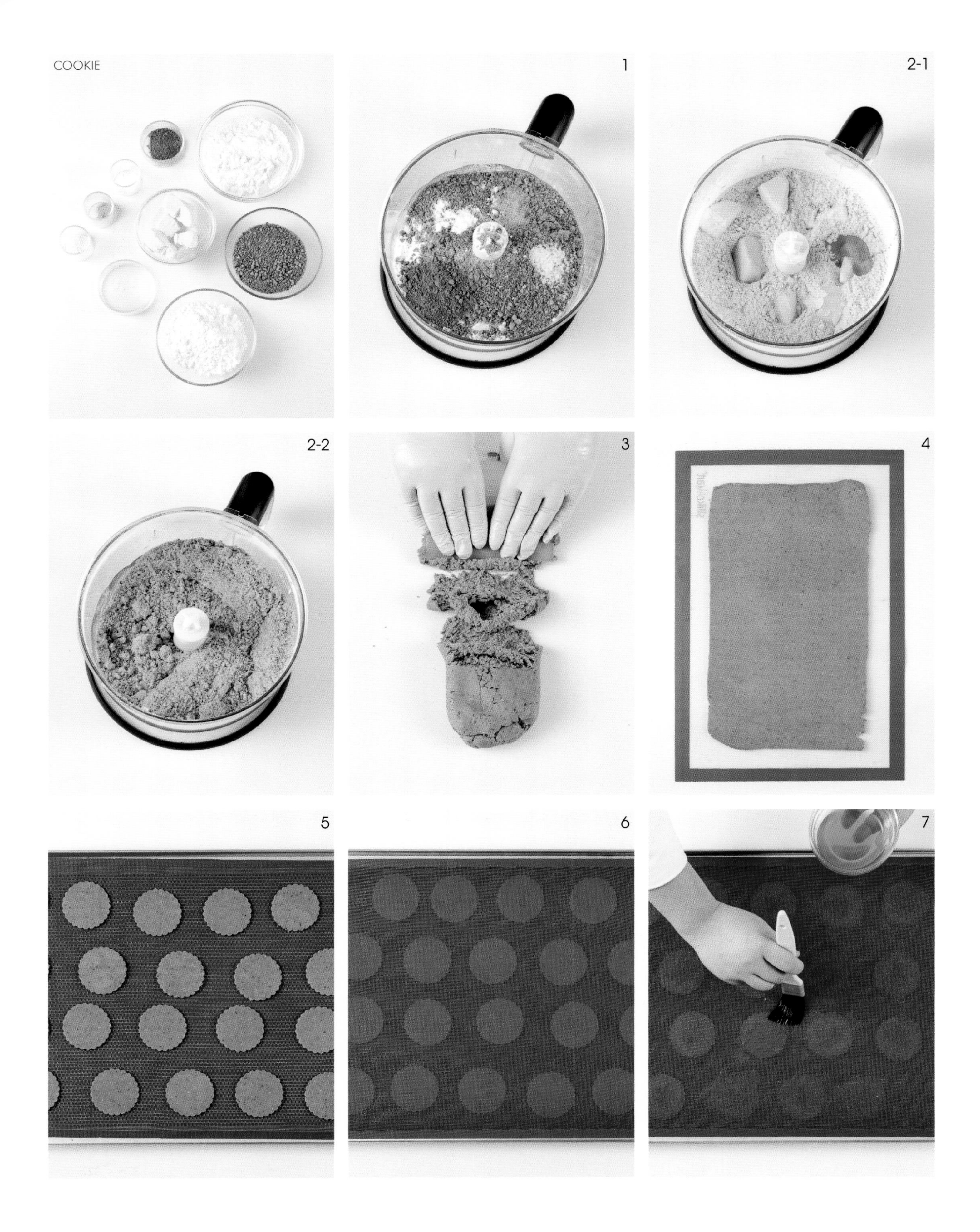
COOKIE
1
2-1
2-2
3
4
5
6
7

Process

쿠키

1. 푸드프로세서에 버터와 달걀흰자를 제외한 모든 재료를 넣고 혼합한다.
2. 버터, 달걀흰자를 넣고 반죽이 뭉치기 시작할 때까지 혼합한다.
   * 버터는 20℃ 정도의 상태로 준비한다.
3. 푸드프로세서에서 꺼낸 반죽을 스크래퍼로 이겨 매끄럽게 혼합한다.
4. 3mm 두께로 밀어 편 후 냉동고에서 단단하게 굳힌다.
5. 6.5cm 국화 모양 커터로 커팅한 후 두 장의 에어 매트 사이에 팬닝한다.
6. 160℃ 오븐에서 15분간 굽는다.
7. 구워져 나온 직후 녹인 카카오버터(분량 외)를 발라준다.

COOKIE

1. Combine all the ingredients, except butter and egg whites, in a food processor.
2. Add butter and egg whites and mix until the dough starts to come together.
   * Prepare butter to about 20°C.
3. Remove the dough from the food processor. Knead it with a scraper to make a smooth mixture.
4. Roll out to 3 mm and set it cold in a freezer.
5. Cut out with a 6.5 cm round crinkled cutter and place them between two perforated mats.
6. Bake for 15 minutes at 160°C.
7. Brush with melted cacao butter (other than requested) immediately after baking.

PISTACHIO BUTTER GANACHE
8
9
10
11
12
ORANGE FILLING
13
14
15
16

피스타치오 버터 가나슈

8. 볼에 포마드 상태의 버터, 전화당을 넣고 부드럽게 풀어준다.
9. 피스타치오 페이스트를 넣고 혼합한다.
10. 구운 피스타치오 분태를 넣고 혼합한다.
11. 녹인 화이트초콜릿(30℃)을 조금씩 나눠 넣어가며 혼합한다.
12. 오렌지 리큐어를 넣고 혼합한다.

오렌지 충전물

13. 오렌지는 끓는 물에 넣고 껍질이 완전히 부드러워질 때까지 약 1시간 30분간 삶는다.
    * 이때 15분마다 새 물로 교체한다.
14. 물기를 제거한 후 잘게 자른다.
15. NH펙틴과 설탕을 섞어준다.
16. 냄비에 **14**의 전처리한 오렌지 165g, 오렌지주스, 패션푸르트 퓌레, 레몬주스, 꿀, 트레할로스를 넣고 가열하다가 40℃가 되면 **15**를 넣고 혼합한다.

PISTACHIO BUTTER GANACHE

8. Soften room-temperature butter and inverted sugar in a bowl.
9. Mix with pistachio paste.
10. Add chopped roasted pistachios and mix.
11. Gradually add melted white chocolate (30°C) little by little to combine.
12. Add orange liqueur and mix.

ORANGE FILLING

13. Blanch orange in boiling water until the skin becomes completely soft, about 1 hour and 30 minutes.
    * Replace the water every 15 minutes.
14. Drain water and chop finely.
15. Mix pectin NH with sugar.
16. Heat 165 grams of pre-treated orange from **14**, orange juice, passion fruit puree, lemon juice, honey, and trehalose in a pot. When it's 40°C, combine with **15**.

17
18
19
20
FINISH
21
PISTACHIO
22
23
24
25
26
PISTACHIO

17. 45Brix가 될 때까지 가열한다.
18. 핸드블렌더로 곱게 갈아준다.
19. 트레이에 부어 차갑게 식힌다.
20. 부드럽게 풀어 사용한다.

마무리

21. 314-315p와 동일한 방법으로 초콜릿 장식물을 준비한다.
22. 완전히 식힌 두 장의 쿠키 사이에 피스타치오 버터 가나슈를 파이핑한다.
23. 피스타치오 버터 가나슈 중심부에 오렌지 충전물을 파이핑한다.
24. 샌딩한 후 수평이 되도록 가볍게 눌러준다.
25. 여분의 초콜릿을 소량 파이핑한다.
26. 장식물을 올려 고정시킨다.

17. Heat until it reaches 45 Brix.
18. Finely grind with an immersion blender.
19. Pour onto a tray and cool completely.
20. Soften to use.

FINISH

21. Prepare the chocolate decorations the same on pages 314-315.
22. Pipe pistachio butter ganache between two pieces of cooled cookies.
23. Pipe orange filling in the center of the pistachio butter ganache.
24. Press lightly to level the cookies.
25. Pipe a small amount of chocolate on the cookie.
26. Attach the decoration.

PISTACHIO

RASPBERRY
SPICE
SPICE
RASPBERRY
PISTACHIO
SPICE
SPICE
SPICE
PISTACHIO

BUTTER
CARAMEL

[39] SALTED BUTTER CARAMEL COOKIE

39 SALTED BUTTER CARAMEL COOKIE

솔티드 버터 캐러멜 쿠키

ingredients - Ø 6.5cm, 24 cookies

GLUTEN FREE

캐러멜 파우더*

설탕 150g

쿠키

캐러멜 파우더* 89g
게랑드소금 2.6g
아몬드파우더 89g
고운 쌀가루 149g
베이킹파우더 0.6g
버터 109g
달걀흰자 12g

솔티드 캐러멜 충전물

생크림 224g
게랑드소금 2.4g
바닐라빈 1/2개
(마다가스카르산)
물엿 147g
설탕 147g
젤라틴매스 7.1g
버터 73g

기타

블론드초콜릿
(DULCEY 35%)
게랑드소금

GLUTEN FREE

CARAMEL POWDER*

150g Sugar

COOKIE

89g Caramel powder*
2.6g Guérande salt
89g Almond powder
149g Fine rice flour
0.6g Baking powder
109g Butter
12g Egg whites

SALTED CARAMEL FILLING

224g Heavy cream
2.4g Guérande salt
1/2 Vanilla bean
(Madagascar)
147g Corn syrup
147g Sugar
7.1g Gelatin mass
73g Butter

OTHER

Blonde chocolate
(DULCEY 35%)
Guérande salt

CARAMEL POWDER
1
2
3
4
COOKIE
5
6
7
8
9
10
11

Process

캐러멜 파우더

1. 냄비에 설탕을 넣고 가열해 짙은 갈색의 캐러멜을 만든다.
2. 완성한 캐러멜을 실팻에 붓고 완전히 식힌다.
3. 푸드프로세서에 넣고 갈아 고운 파우더를 만든다.
4. 완성된 캐러멜 파우더는 습기에 취약하므로 밀봉해 보관한다.

쿠키

5. 푸드프로세서에 버터와 달걀흰자를 제외한 모든 재료를 넣고 혼합한다.
6. 버터, 달걀흰자를 넣고 반죽이 뭉치기 시작할 때까지 혼합한다.
   * 버터는 20℃ 정도의 상태로 준비한다.
7. 푸드프로세서에서 꺼낸 반죽을 스크래퍼로 이겨 매끄럽게 혼합한다.
8. 3mm 두께로 밀어 편 후 냉동고에서 단단하게 굳힌다.
9. 6.5cm 국화 모양 커터로 커팅한 후 두 장의 에어 매트 사이에 팬닝한다.
10. 160℃ 오븐에서 15분간 굽는다.
11. 구워져 나온 직후 녹인 카카오버터(분량 외)를 발라준다.

CARAMEL POWDER

1. Heat sugar in a pot to make dark brown caramel.
2. Pour the caramel over a Silpat and let cool completely.
3. Grind in a food processor to make it into a fine powder.
4. Since the caramel powder is vulnerable to moisture, store it sealed.

COOKIE

5. Combine all the ingredients, except butter and egg whites, in a food processor.
6. Add butter and egg whites and mix until the dough starts to come together.
   * Prepare butter to about 20°C.
7. Remove the dough from the food processor. Knead it with a scraper to make a smooth mixture.
8. Roll out to 3 mm and set it cold in a freezer.
9. Cut out with a 6.5 cm round crinkled cutter and place them between two perforated mats.
10. Bake for 15 minutes at 160°C.
11. Brush with melted cacao butter (other than requested) immediately after baking.

SALTED CARAMEL FILLING
12
13
14
15
16
17
18
19

솔티드 캐러멜 충전물

12. 냄비에 생크림, 게랑드소금, 바닐라빈을 넣고 끓기 직전까지 가열한다.
13. 다른 냄비에 물엿을 넣고 설탕을 조금씩 나눠 넣어가며 캐러멜화시킨다.
14. **12**를 조금씩 나눠 넣어가며 디글레이즈한다.
15. 108℃까지 가열한다.
16. 체에 거른다.
17. 80℃까지 식힌 후 젤라틴매스를 넣고 섞어준다.
18. 차가운 상태의 버터를 넣고 핸드블렌더로 혼합한다.
19. 실온에서 12시간 휴지시킨 후 사용한다.

SALTED CARAMEL FILLING

12. Heat heavy cream, Guérande salt, and vanilla bean until just before it starts to boil.
13. In a separate bowl, heat corn syrup and gradually add sugar little by little to caramelize.
14. Add **12** a little bit at a time to deglaze.
15. Heat to 108°C.
16. Filter through a sieve.
17. Cool to 80°C and add gelatin mass.
18. Add cold butter and combine with an immersion blender.
19. Let stand at an ambient temperature for 12 hours before use.

마무리

20. 진공 성형기를 이용해 초콜릿 몰드를 만든다. (356p)
템퍼링한 블론드초콜릿으로 몰딩해 굳힌 후 솔티드 버터 캐러멜 충전물을 가득 채운다.

21. 게랑드소금을 토핑한다.

22. 완전히 식힌 쿠키를 솔티드 버터 캐러멜 충전물 위에 올려 고정시킨다.

23. 냉장고에 10분 정도 두어 초콜릿을 충분히 수축시킨 후 몰드와 분리한다.

FINISH

20. Make a chocolate mold using a vacuum former (p.356).
    Coat the mold with tempered blonde chocolate. Fill it with salted butter caramel filling after the chocolate sets.
21. Sprinkle Guérande salt.
22. Place and fix the completely cooled cookie over the salted butter caramel filling.
23. Refrigerate for about 10 minutes to sufficiently crystallize the chocolate, then separate it from the mold.

PRALINE
PRALINE

40 PRALINE COOKIE

프랄리네 쿠키

ingredients - Ø 6.5cm, 24 cookies

쿠키	헤이즐넛 누가틴*	헤이즐넛 잔두야	기타
슈거파우더 44g	버터 23g	밀크초콜릿 122g (JIVARA LATTE 40%)	밀크초콜릿 (JIVARA LATTE 40%)
무스코바도 설탕 44g	물엿 16g	헤이즐넛 프랄리네(50p) 107g (수제)	
게랑드소금 2.6g	게랑드소금 0.7g	헤이즐넛 누가틴* 35g	
아몬드파우더 89g	NH펙틴 0.8g	에클라도르 35g (ECLAT D'OR)	
고운 쌀가루 149g	설탕 64g	게랑드소금 2.3g	
베이킹파우더 0.6g	헤이즐넛 분태 96g		
버터 110g			
달걀흰자 12g			

COOKIE	HAZELNUT NOUGATINE*	HAZELNUT GIANDUJA	OTHER
44g Powdered sugar	23g Butter	122g Milk chocolate (JIVARA LATTE 40%)	Milk chocolate (JIVARA LATTE 40%)
44g Muscovado sugar	16g Corn syrup	107g Hazelnut praliné (p.50) (handcrafted)	
2.6g Guérande salt	0.7g Guérande salt	35g Hazelnut nougatine*	
89g Almond powder	0.8g Pectin NH	35g Eclat d'Or (ECLAT D'OR)	
149g Fine rice flour	64g Sugar	2.3g Guérande salt	
0.6g Baking powder	96g Chopped hazelnuts		
110g Butter			
12g Egg whites			

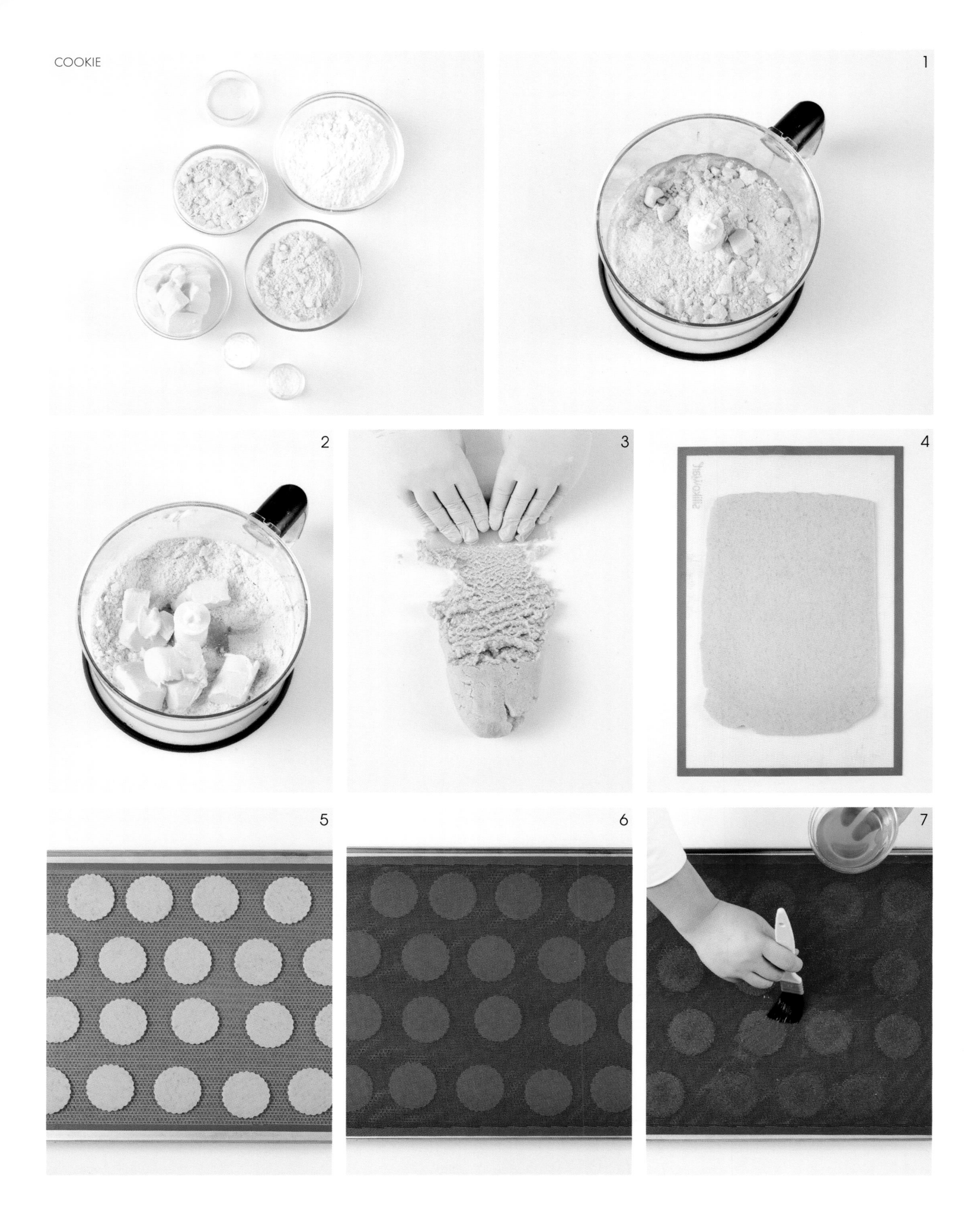
1
2
3
4
5
6
7

Process

쿠키

1. 푸드프로세서에 버터와 달걀흰자를 제외한 모든 재료를 넣고 혼합한다.
2. 버터, 달걀흰자를 넣고 반죽이 뭉치기 시작할 때까지 혼합한다.
   * 버터는 20℃ 정도의 상태로 준비한다.
3. 푸드프로세서에서 꺼낸 반죽을 스크래퍼로 이겨 매끄럽게 혼합한다.
4. 3mm 두께로 밀어 편 후 냉동고에서 단단하게 굳힌다.
5. 6.5cm 국화 모양 커터로 커팅한 후 두 장의 에어 매트 사이에 팬닝한다.
6. 160℃ 오븐에서 15분간 굽는다.
7. 구워져 나온 직후 녹인 카카오버터(분량 외)를 발라준다.

COOKIE

1. Combine all the ingredients, except butter and egg whites, in a food processor.
2. Add butter and egg whites and mix until the dough starts to come together.
   * Prepare butter to about 20℃.
3. Remove the dough from the food processor. Knead it with a scraper to make a smooth mixture.
4. Roll out to 3 mm and set it cold in a freezer.
5. Cut out with a 6.5 cm round crinkled cutter and place them between two perforated mats.
6. Bake for 15 minutes at 160℃.
7. Brush with melted cacao butter (other than requested) immediately after baking.

HAZELNUT NOUGATINE
8
9
10
11
12
HAZELNUT GIANDUJA
13
14

헤이즐넛 누가틴

8. 냄비에 버터, 물엿, 게랑드소금을 넣고 가열한다.
9. 버터가 완전히 녹으면 미리 섞어둔 NH펙틴과 설탕을 넣고 끓을 때까지 가열한다.
10. 헤이즐넛 분태를 넣고 혼합한다.
    * 헤이즐넛 분태는 150℃ 오븐에서 5분간 구워 사용한다.
11. 실팻에 얇게 펼친다.
12. 160℃ 오븐에서 17~20분간 구운 후 식혀 조각내 사용한다.

헤이즐넛 잔두야

13. 볼에 녹인 밀크초콜릿, 헤이즐넛 프랄리네, 헤이즐넛 누가틴, 에클라도르, 게랑드소금을 넣고 혼합한다.
14. 26℃로 템퍼링한 후 사용한다.

HAZELNUT NOUGATINE

8. Heat butter, corn syrup, and Guérande salt in a pot.
9. When the butter is completely melted, add previously mixed pectin NH and sugar, and continue to heat until it boils.
10. Stir in chopped hazelnuts.
    * Bake chopped hazelnuts for 5 minutes at 150°C to use.
11. Spread thinly over a Silpat.
12. Bake for 17~20 minutes at 160°C. Cool completely, and coarsely chop to use.

HAZELNUT GIANDUJA

13. Combine melted milk chocolate, hazelnut praliné, hazelnut nougatine, eclat d'Or, and Guérande salt in a bowl.
14. Temper to 26°C to use.

마무리

15. 진공 성형기를 이용해 초콜릿 몰드를 만든다. (356p)
    템퍼링한 밀크초콜릿으로 몰딩해 굳힌 후 헤이즐넛 잔두야를 가득 채운다.
16. 완전히 식힌 쿠키를 헤이즐넛 잔두야 위에 올려 고정시킨다.
17. 냉장고에 10분 정도 두어 초콜릿을 충분히 수축시킨 후 몰드와 분리한다.

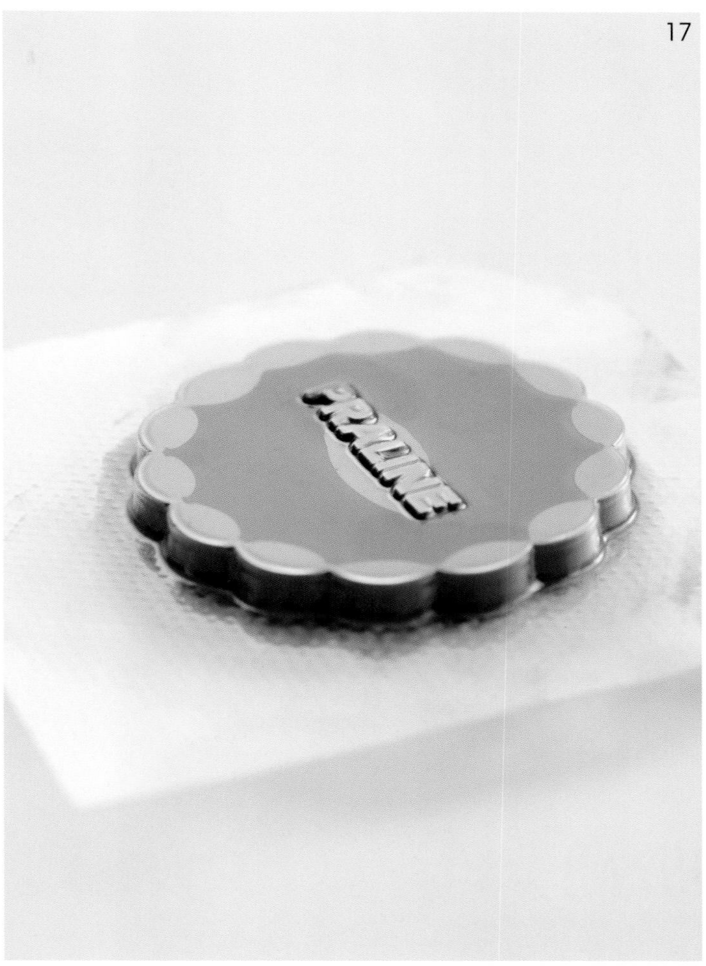

FINISH

15. Make a chocolate mold using a vacuum former (p.356).
Coat the mold with tempered milk chocolate, then fill with hazelnut gianduja after the chocolate sets.

16. Place and fix the completely cooled cookie over the hazelnut gianduja.

17. Refrigerate for about 10 minutes to sufficiently crystallize the chocolate, then separate it from the mold.

100%
CHOCOLATE

41 # 100% CHOCOLATE COOKIE

100% 초콜릿 쿠키

ingredients - Ø 6.5cm, 24 cookies

쿠키

슈거파우더 90g
게랑드소금 2.6g
아몬드파우더 90g
고운 쌀가루 128g
카카오파우더 17g
베이킹파우더 0.6g
버터 110g
달걀흰자 12g

카카오닙 누가틴*

버터 23g
물엿 16g
게랑드소금 0.7g
NH펙틴 0.8g
설탕 64g
카카오닙 96g

카카오 잔두야

카카오닙 누가틴* 26g
다크초콜릿 100g
(GUANAJA 70%)
헤이즐넛 프랄리네(50p) 146g
(수제)
에클라도르 26g
(ECLAT D'OR)
게랑드소금 1.3g

기타

다크초콜릿
(GUANAJA 70%)

COOKIE

90g Powdered sugar
2.6g Guérande salt
90g Almond powder
128g Fine rice flour
17g Cacao powder
0.6g Baking powder
110g Butter
12g Egg whites

CACAO NIBS NOUGATINE*

23g Butter
16g Corn syrup
0.7g Guérande salt
0.8g Pectin NH
64g Sugar
96g Cacao nibs

CACAO GIANDUJA

26g Cacao nibs nougatine*
100g Dark chocolate
(GUANAJA 70%)
146g Hazelnut praliné (p.50)
(handcrafted)
26g Eclat d'Or
(ECLAT D'OR)
1.3g Guérande salt

OTHER

Dark chocolate
(GUANAJA 70%)

COOKIE
1
2-1
2-2
3
4
5
6
7

Process

쿠키

1. 푸드프로세서에 버터와 달걀흰자를 제외한 모든 재료를 넣고 혼합한다.
2. 버터, 달걀흰자를 넣고 반죽이 뭉치기 시작할 때까지 혼합한다.
   * 버터는 20℃ 정도의 상태로 준비한다.
3. 푸드프로세서에서 꺼낸 반죽을 스크래퍼로 이겨 매끄럽게 혼합한다.
4. 3mm 두께로 밀어 편 후 냉동고에서 단단하게 굳힌다.
5. 6.5cm 국화 모양 커터로 커팅한 후 두 장의 에어 매트 사이에 팬닝한다.
6. 160℃ 오븐에서 15분간 굽는다.
7. 구워져 나온 직후 녹인 카카오버터(분량 외)를 발라준다.

COOKIE

1. Combine all the ingredients, except butter and egg whites, in a food processor.
2. Add butter and egg whites and mix until the dough starts to come together.
   * Prepare butter to about 20℃.
3. Remove the dough from the food processor. Knead it with a scraper to make a smooth mixture.
4. Roll out to 3 mm and set it cold in a freezer.
5. Cut out with a 6.5 cm round crinkled cutter and place them between two perforated mats.
6. Bake for 15 minutes at 160℃.
7. Brush with melted cacao butter (other than requested) immediately after baking.

CACAO NIBS NOUGATINE
8
9
10
11
12
CACAO GIANDUJA
13-1
13-2
14
15

카카오닙 누가틴

8. 냄비에 버터, 물엿, 게랑드소금을 넣고 가열한다.
9. 버터가 완전히 녹으면 미리 섞어둔 NH펙틴과 설탕을 넣고 끓을 때까지 가열한다.
10. 카카오닙을 넣고 혼합한다.
11. 실팻에 얇게 펼친다.
12. 160℃ 오븐에서 22~24분간 구운 후 완전히 식혀 조각내 사용한다.

카카오 잔두야

13. 분쇄기에 카카오닙 누가틴을 넣고 곱게 간다.
14. 녹인 다크초콜릿, 헤이즐넛 프랄리네, 에클라도르, 게랑드소금을 넣고 혼합한다.
15. 26℃로 템퍼링한 후 사용한다.

CACAO NIBS NOUGATINE

8. Heat butter, corn syrup, and Guérande salt in a pot.
9. When the butter is completely melted, add previously mixed pectin NH and sugar, and continue to heat until it boils.
10. Stir in cacao nibs.
11. Spread thinly over a Silpat.
12. Bake for 22~24 minutes at 160°C. Cool completely, and coarsely chop to use.

CACAO GIANDUJA

13. Finely grind cacao nibs nougatine in a grinder.
14. Combine melted dark chocolate, hazelnut praliné, eclat d'Or, and Guérande salt.
15. Temper to 26°C to use.

DESIGN FILM
16
17
18
19
20
FINISH
21
22
23

디자인 필름

16. 진공 성형기를 이용해 초콜릿 몰드를 만든다.
17. 완성된 몰드에 템퍼링한 다크초콜릿을 가득 채워준다.
18. 몰드를 뒤집어 여분의 초콜릿을 털어낸다.
19. 스크래퍼를 이용해 몰드를 정리한다.
20. 초콜릿을 완전히 굳힌다.

마무리

21. 초콜릿이 충분히 굳으면 완성한 카카오 잔두야를 가득 채워준다.
22. 완전히 식힌 쿠키를 카카오 잔두야 위에 올려 고정시킨다.
23. 냉장고에 10분 정도 두어 초콜릿을 충분히 수축시킨 후 몰드와 분리한다.

DESIGN FILM

16. Make a chocolate mold using a vacuum former.
17. Fill with tempered dark chocolate in the mold.
18. Turn over the mold and shake to remove excess chocolate.
19. Organize the mold with a scraper.
20. Let the chocolate set completely.

FINISH

21. Fill with cacao gianduja after the chocolate has set.
22. Place and fix the completely cooled cookie over the gianduja.
23. Refrigerate for about 10 minutes to sufficiently crystallize the chocolate, then separate it from the mold.

PATISSERIE
by
GARUHARU

TASTE

계절에 따른 가장 좋은 재료를 탐구합니다. 이렇게 찾은 재료의 맛이 디저트에 충분히 표현되었는지, 맛의 밸런스가 조화로운지 확인합니다.

We explore the best ingredient for each season. We make sure that the taste of the ingredients found is sufficiently expressed in the dessert and that the balance of the taste is harmonious.

TEXTURE

테크닉적으로 좋은 텍스처를 완성하는 것에 더해 하나의 디저트 안에서 단조롭지 않은 다양한 텍스처의 재미를 느낄 수 있도록 구성합니다.

In addition to mastering a technically good texture, we try to compose to experience the fun in various textures in one dessert that will not be monotonous.

DESIGN

디저트의 맛을 연상시킬 수 있는 포인트를 담은 간결한 디자인을 추구합니다.

We pursue a simple design with an emphasis that resembles the taste of the dessert.

100%
CHOCOLATE
SALTED BUTTER
CARAMEL
PRALINE

Team GARUHARU

가루하루의 시작부터 지금까지 새로운 시도와 도전에 늘 열정적으로 동참해주는 가루하루 팀에게 고마운 마음을 전합니다.
성실하고 열정적인 재능 있는 동료들과 함께여서 새로운 시도를 주저하지 않고 무모했던 도전의 과정을 즐길 수 있었습니다.

I would like to express my gratitude to Team GARUHARU for their passionate participation in new attempts and challenges since the beginning of GARUHARU. Team GARUHARU for their passionate participation in new attempts and challenges. With these talented teammates who are sincere and enthusiastic, I was able to enjoy the process of reckless challenges without hesitating to try new things.